T0222815

SQL Server Data Automation Through Frameworks

Building Metadata-Driven Frameworks with T-SQL, SSIS, and Azure Data Factory

Andy Leonard
Kent Bradshaw

Apress®

SQL Server Data Automation Through Frameworks: Building Metadata-Driven Frameworks with T-SQL, SSIS, and Azure Data Factory

Andy Leonard
Farmville, VA, USA

Kent Bradshaw
Providence Forge, VA, USA

ISBN-13 (pbk): 978-1-4842-6212-2
https://doi.org/10.1007/978-1-4842-6213-9

ISBN-13 (electronic): 978-1-4842-6213-9

Managing Director, Apress Media LLC: Welmoed Spahr
Acquisitions Editor: Jonathan Gennick
Development Editor: Laura Berendson
Coordinating Editor: Jill Balzano

Cover image designed by Freepik (www.freepik.com)

Distributed to the book trade worldwide by Springer Science+Business Media New York, 233 Spring Street, 6th Floor, New York, NY 10013. Phone 1-800-SPRINGER, fax (201) 348-4505, e-mail orders-ny@springer-sbm.com, or visit www.springeronline.com. Apress Media, LLC is a California LLC and the sole member (owner) is Springer Science + Business Media Finance Inc (SSBM Finance Inc). SSBM Finance Inc is a **Delaware** corporation.

For information on translations, please e-mail booktranslations@springernature.com; for reprint, paperback, or audio rights, please e-mail bookpermissions@springernature.com.

Apress titles may be purchased in bulk for academic, corporate, or promotional use. eBook versions and licenses are also available for most titles. For more information, reference our Print and eBook Bulk Sales web page at http://www.apress.com/bulk-sales.

Any source code or other supplementary material referenced by the author in this book is available to readers on GitHub via the book's product page, located at www.apress.com/9781484262122. For more detailed information, please visit http://www.apress.com/source-code.

Printed on acid-free paper

For Christy
—Andy

For Ann
—Kent

Table of Contents

About the Authors

Andy Leonard is Chief Data Engineer at Enterprise Data & Analytics, creator and Data Philosopher at DILM (Data Integration Lifecycle Management) Suite, an Azure Data Factory and SQL Server Integration Services trainer and consultant, and a BimlHero. He is a SQL Server database and data warehouse developer, community mentor, engineer, and farmer. Andy is coauthor of *SQL Server Integration Services Design Patterns*, *Data Integration Life Cycle Management with SSIS*, and *The Biml Book*.

Kent Bradshaw is the founder or Tudor Data Solutions, LLC. With over 40 years of IT experience, he is a SQL Server database/ETL developer and database architect with a background in Medicaid claims, public schools, government, retail, and insurance systems. In 2011, Kent founded Tudor Data Solutions, LLC, to pursue new development opportunities which led to his association with Andy Leonard and Enterprise Data & Analytics. In 2017, Kent received the MPP certification for Data Science.

About the Technical Reviewer

 **André van Meulebrouck** has a keen interest in functional programming, especially Haskell and F#.

He also likes data technologies from markup languages to databases and F# type providers.

He lives in Southern California with his wife "Tweety", and is active in athletics: hiking, mountain biking, and gravity/balance sports like freestyle skating (in-line and ice), skateboarding, surfing, and sandboarding.

To keep his mind sharp, he does compositional origami, plays classical guitar, and enjoys music notation software.

Acknowledgments

This book would not have been possible without the help of Kent Bradshaw, my coauthor, friend, and brother. This book represents Kent's first foray into authoring, and he delivered outstanding work.

Thanks André van Meulebrouck for proofreading, working through the samples, and providing feedback on the manuscript.

I owe my coworkers at Enterprise Data & Analytics (EntDNA.com) a debt of gratitude for their encouragement and for covering for me when I stayed up too late trying to figure out how to make the code work. We have an awesome team, especially my brothers from other mothers – Nick Harris, Shannon Lowder, and Reeves Smith.

Donald Farmer inspires me every time we interact. As Principal Program Manager at Microsoft, Donald worked extensively with SSIS and helped shape the product. Donald continues to shape data software by providing vendors unique strategic guidance at TreeHive Strategy (treehivestrategy.com).

I am certain there are many excellent editors in this business. Jonathan Gennick is the best with whom I have had the privilege to work. Without Jill Balzano's input and project management, I would have been lost!

Finally, I thank my family for their understanding. My children Stevie Ray, Emma, and Riley who live at home at the time of this writing, and Manda and Penny who have children of their own. And Christy, to whom I dedicate this book, my wife, my love. Thank you.

—Andy

Writing this book never really entered my mind until my great friend and EDNA colleague, Andy Leonard, convinced me to do it. I truly appreciate his confidence in me, and it was a very interesting experience. I'm really glad that I could work together with him on this project.

Many thanks to Jonathan Gennick, Jill Balzano, and André van Meulebrouck for all of their help and suggestions. They made things so much easier.

Most important of all, I want to thank my wife, Ann. Much of where I am today is due to her love, support, and encouragement. I couldn't begin to do it without her. Thank you.

—Kent

Introduction

Frameworks have existed for as long as people have been building things. When people began developing solutions using computers, frameworks soon followed. Like woodworking jigs, frameworks exist to simplify work and speed up the process of developing a solution, whether the problem is to complete a woodworking project or deliver an enterprise data integration solution.

Who Is This Book For?

Anyone who is interested in data-related automation or executing multiple chunks of code with a single command will find value in these pages. Three audiences will particularly benefit from this book:

1. Database developers

2. Data engineers and data integration developers

3. Business Intelligence Markup Language (Biml) developers

Database developers will learn about driving database object execution from stored procedures and will see one example of a database framework.

Data engineers and data integration developers will learn how to use SQL Server Integration Services (SSIS) to implement a metadata-driven data integration framework.

Biml developers will see one example of using BimlExpress and metadata to rapidly produce several SSIS packages in selected design patterns *and* an example of interrogating flat files to generate flat file formats and flat file connection managers via Biml.

How This Book Is Structured

In the following sections, we will explore the chapter layout of this book. There are two parts that guide you through the process of building a database framework using stored procedures and an SSIS framework. Each of those parts is composed of several chapters.

Part I: Stored Procedure–Based Database Frameworks

In the first part of this book, we will focus on building a database framework using stored procedures. These chapters are targeted primarily at those who develop database solutions. We do not endeavor to create the most comprehensive primer for every aspect of database development; we cover the basics well.

Chapter 1: Stored Procedures 101

Before we dive into database development, let's cover some basics of stored procedures.

Chapter 2: Automation with Stored Procedures

Once you have a solid grasp on stored procedures, we will use what we've learned about stored procedures to automate execution using the controller pattern, which executes one or more stored procedures.

Chapter 3: Stored Procedure Orchestrators

In this chapter, we examine high-level execution management using the orchestrator pattern, which executes one or more controller stored procedures.

Chapter 4: A Stored Procedure–Based Metadata-Driven Framework

Metadata-driven frameworks are a recurring theme in this book. In this chapter, we store information about stored procedures, controllers, and orchestrators to execute framework applications.

Part II: SSIS Frameworks

If your enterprise is executing a few SSIS packages, there is no need for an SSIS framework. However, if your enterprise is executing several hundred (or several thousand) SSIS packages, you *definitely* need an SSIS framework.

Chapter 5: A Simple, Custom, File-Based SSIS Framework

In this chapter, we define a data integration/engineering framework metadata database that manages configuration and execution. We introduce a version of an SSIS framework that manages these subject areas for SSIS executed on-premises.

Chapter 6: Framework Execution Engine

In this chapter, we add instrumentation in the form of Information and Error events. Built-in SSIS logging will surface these messages, which are useful when troubleshooting SSIS framework application issues.

Chapter 7: Framework Logging

In this chapter, we persist execution metadata to "Instance" tables. The ApplicationInstance table stores one record for each instance of an SSIS framework application execution. The ApplicationPackageInstance table stores one record for each instance of an SSIS framework application package execution. Important execution instance attributes – such as start and end times and execution status – are maintained in the "Instance" tables.

Chapter 8: Azure-SSIS Integration Runtime

In this chapter, we introduce Azure Data Factory (ADF) and the Azure-SSIS integration runtime by walking through the provisioning processes.

Chapter 9: Deploy a Simple, Custom, File-Based Azure-SSIS Framework

In Chapters 5–7, we designed and constructed an SSIS framework aimed at on-premises SSIS execution, configuration, and logging. In this chapter, we begin migrating the SSIS framework designed in Chapters 5–7 by provisioning a new Azure SQL database and then deploying the metadata database to our new Azure SQL database. We provision an Azure File Share and migrate test SSIS packages to the Azure File Share. Finally, we begin building the Azure Data Factory parent pipeline that serves as the execution engine for the ADF version of the SSIS framework.

Chapter 10: Framework Logging in ADF

In this chapter, we add logging functionality to the parent ADF pipeline, much like the functionality we added to the on-premises version of the SSIS framework in Chapter 7.

Chapter 11: Fault Tolerance in the ADF Framework

In this chapter, we complete ADF execution engine functionality by implementing fault tolerance to programmatically stop (or not stop) pipeline execution based on SSISConfig metadata configurations.

Conclusion

This book is for people who want to dive deeper into SQL Server and SSIS automation. We will be discussing and demonstrating database and Integration Services frameworks and covering such topics as SQL Server, SSIS, Azure, and Azure Data Factory.

As you can see, we have a long journey filled with learning and new insight ahead of us. Let's get started!

PART I

Stored Procedure-Based Database Frameworks

CHAPTER 1

Stored Procedures 101

One of the most common issues continually facing IT organizations is finding the proper balance between the effort to develop and deploy processes into production against the efficiency and effectiveness of production control operators. The effort to develop and deploy and the effectiveness of production control seem to be diametrically opposed. Making it easier to develop and deploy processes usually means more work and manual intervention for production control. The real question that needs to be considered is where is it better to "feel the pain"? Pushing the effort toward the development side of the equation can slow down the throughput but minimizes the liability of issues at the production process level. This and the next few chapters are going to concentrate on processes that are executed using stored procedures. In this chapter, you'll get a basic introduction to stored procedures that is the foundation for the chapters that follow. You'll see how to create a child stored procedure, and we'll provide you a template that you can use to create similar procedures in your own work.

The Need for a Framework

When just in development mode, it is awfully easy (and impressive) to construct large, monolithic procedures that do everything from beginning to end. They are the proverbial "black box" where something goes in, many gyrations take place, and then the desired result occurs. Those are great until a problem occurs, or just a change in business requirements means that modifications need to be made. When the procedure does so much, what does it take to test it once the modifications are done? Even though the change only impacts 10 percent of the procedure, the entire procedure has to be tested. What does it take to accomplish that? What if there are several intricate modifications that need to be made? How difficult does it become for more than one developer to work on the changes and coordinate their efforts?

3

© Andy Leonard, Kent Bradshaw 2020
A. Leonard and K. Bradshaw, *SQL Server Data Automation Through Frameworks*,
https://doi.org/10.1007/978-1-4842-6213-9_1

Now, consider that monolith broken up into multiple procedures, each of which performs a unit of work. When a change is made and needs to be tested, that effort can be isolated to just what is necessary to perform that unit of work, and the validation is concentrated on the result of the procedure. And, with it now being multiple procedures, modifications can usually be done simultaneously by multiple developers, and their efforts can be mutually exclusive. Over time, that approach can prove to be much more cost effective and efficient.

That is where a framework helps to organize and manage processes to provide the most flexibility in development and can minimize the maintenance effort (which, sometimes, is not considered until it becomes an obvious issue). A framework provides a consistent methodology for assembling and executing processes. It also promotes writing code in small units of work that can potentially be mixed, matched, and reused. It adds complexity to the development and deployment processes but can reduce the effort for production scheduling. The framework can also provide greater flexibility for managing the execution of the process.

Demonstration of a Framework

To begin the analysis of the framework concept, we need a process. Our example to follow shows a framework built to run a daily process against an example schema. The details of that process don't matter to the example. Just consider that any production system might have something that needs to be done each day, and what follows is a framework by which to make those daily processes happen.

Also, part of the example is a monthly process. Just as a system might need certain tasks to be done each day, it's also common to have certain things that need to be done once monthly. In designing such a system, one must take into account the order in which daily and monthly processes execute when their schedules intersect on – in our example – the first of each month.

For the purpose of this book, a simple process has been developed (NOTE: all of the code described can be downloaded at entdna.com. You can also find a link to the code from the book's catalog page on Apress.com). Downloading the example code enables you to follow along with the upcoming examples on your own machine.

An Example Schema

Listing 1-1 shows code to create a schema called FWDemo that will contain everything needed for the demonstration. Also, there is code to create a table called FWDemo. ProcessLog. Including a pattern for writing to this table throughout all of the procedures certainly adds some complexity and overhead to the procedures, but what it provides in monitoring and troubleshooting more than makes up for the upfront effort.

Listing 1-1. Schema and log table creation

```
print 'FWDemo Schema'
If Not Exists(Select name
              From sys.schemas
              Where name='FWDemo')
begin
  print ' - Creating FWDemo schema'
  declare @sql varchar(255) = 'Create Schema FWDemo'
  exec(@sql)
  print ' - FWDemo schema created'
end
Else
print ' - FWDemo schema already exists.'
print ''
GO

IF  EXISTS (SELECT * FROM sys.objects
            WHERE object_id = OBJECT_ID(N'FWDemo.ProcessLog')
             AND type in (N'U'))
    DROP TABLE FWDemo.ProcessLog
GO

SET ANSI_NULLS ON
GO

SET QUOTED_IDENTIFIER ON
GO

SET ANSI_PADDING ON
GO
```

5

```
CREATE TABLE [FWDemo].[ProcessLog](
      [ProcessLogID]      [int] IDENTITY(1,1) NOT NULL,
      [ProcessLogMessage] [nvarchar](255)     NOT NULL,
      [CreateDate]        [smalldatetime]     NOT NULL
)
GO

SET ANSI_PADDING OFF
GO
```

Do you have a SQL Server instance that you can use for learning purposes? Connect to that instance as an administrator, for example, as the sa user. Then, in SQL Server Management Studio (SSMS), open a "New Query" window, copy the code from Listing 1-1, and execute it to create the example schema used in this and subsequent chapters.

The Daily Process

Listing 1-2 shows the code to create two stored procedures that will make up our demonstration Daily Process. We are providing two procedures in our example because it's common to have more than one, and having two allows us to show how to make the execution of subsequent procedures depend upon the success of earlier ones – because the need to execute a series of procedures and halt or take other actions when an error occurs is the real-life scenario that most of us face.

These procedures (as well as all others that we will be using) can be compiled and executed for your own testing. You will notice that there is some code commented out (lines that are preceded with '--') in each procedure that can be invoked (remove the '--', then recompile) to create an error condition. This ability to create an error condition allows testing for successful and unsuccessful completions that will become more important as we progress through demonstration iterations in later chapters.

For the sake of this exercise, we will declare a business rule for the Daily Process stating that FWDemo.DailyProcess1 must complete successfully before FWDemo. DailyProcess2 can be executed. FWDemo.DailyProcess2 must then complete successfully before the Daily Process can be deemed successfully executed.

Listing 1-2. Daily Process stored procedures

```
If Exists(Select s.name + '.' + p.name
          From sys.procedures p
          Join sys.schemas s
            On s.schema_id = p.schema_id
          Where s.name = 'FWDemo'
            And p.name = 'DailyProcess1')
begin
  print ' - Dropping FWDemo.DailyProcess1 stored procedure'
  Drop Procedure FWDemo.DailyProcess1
  print ' - FWDemo.DailyProcess1 stored procedure dropped'
end
GO

CREATE PROCEDURE FWDemo.DailyProcess1
AS
-------------------------------------------------------------------------------
-------------------------------------------------------------------------------
--
-- Purpose: This procedure is part of the Stored Procedure Framework Demo.
--
-- NOTE: An Error situation can be created for testing/demo purposes by
--     un-commenting the Error code in the body of the procedure.  To return
--     to a procedure with a successful execution, re-comment the code or
--     recompile the original.
--
-------------------------------------------------------------------------------
-------------------------------------------------------------------------------

SET NOCOUNT ON

/*******************************************/
/*  Log the START of the procedure to the process log  */
/*******************************************/
```

```
INSERT INTO FWDemo.ProcessLog (
      ProcessLogMessage,
      CreateDate
)
Values ('Procedure FWDemo.DailyProcess1 - STARTING',
      GETDATE()
)

DECLARE @RetStat int

SET @RetStat = 0

/****************************************/
/*  Force an ERROR CONDITION for this procedure  */
/****************************************/

--INSERT INTO FWDemo.ProcessLog (
--     ProcessLogMessage,
--     CreateDate
--)
--VALUES ('Procedure FWDemo.DailyProcess1 - Problem Encountered',
--     GETDATE()
--)
--SET @RetStat = 1

/*********************************************************/
/*  Log the COMPLETION of the procedure to the process log    */
/*********************************************************/

IF @RetStat = 0
   BEGIN
      INSERT INTO FWDemo.ProcessLog (
            ProcessLogMessage,
            CreateDate
      )
      VALUES ('Procedure FWDemo.DailyProcess1 - COMPLETED',
            GETDATE()
      )
```

```
        END
ELSE
    BEGIN
        INSERT INTO FWDemo.ProcessLog (
                ProcessLogMessage,
                CreateDate
        )
        VALUES ('Procedure FWDemo.DailyProcess1 - ERROR',
        GETDATE()
        )
    END

RETURN @RetStat
GO

If Exists(Select s.name + '.' + p.name
            From sys.procedures p
            Join sys.schemas s
              On s.schema_id = p.schema_id
            Where s.name = 'FWDemo'
              And p.name = 'DailyProcess2')
begin
  print ' - Dropping FWDemo.DailyProcess2 stored procedure'
  Drop Procedure FWDemo.DailyProcess2
  print ' - FWDemo.DailyProcess2 stored procedure dropped'
end
GO

CREATE PROCEDURE FWDemo.DailyProcess2
AS
-----------------------------------------------------------------------------
-----------------------------------------------------------------------------
--
-- Purpose: This procedure is part of the Stored Procedure Framework Demo.
--
```

```
-- NOTE: An Error situation can be created for testing/demo purposes by
--     un-commenting the Error code in the body of the procedure.  To return
--     to a procedure with a successful execution, re-comment the code or
--     recompile the original.
--
-------------------------------------------------------------------------------
-------------------------------------------------------------------------------

SET NOCOUNT ON

/*******************************************/
/*  Log the START of the procedure to the process log  */
/*******************************************/

INSERT INTO FWDemo.ProcessLog (
      ProcessLogMessage,
      CreateDate
)
Values ('Procedure FWDemo.DailyProcess2 - STARTING',
      GETDATE()
)

DECLARE @RetStat int

SET @RetStat = 0

/*****************************************/
/*  Force an ERROR CONDITION for this procedure  */
/*****************************************/

--INSERT INTO FWDemo.ProcessLog (
--     ProcessLogMessage,
--     CreateDate
--)
--VALUES ('Procedure FWDemo.DailyProcess2 - Problem Encountered',
--     GETDATE()
--)
```

```
--SET @RetStat = 1

/*****************************************************/
/*  Log the COMPLETION of the procedure to the process log     */
/*****************************************************/

IF @RetStat = 0
    BEGIN
        INSERT INTO FWDemo.ProcessLog (
                ProcessLogMessage,
                CreateDate
        )
        VALUES ('Procedure FWDemo.DailyProcess2 - COMPLETED',
                GETDATE()
        )
    END
ELSE
    BEGIN
        INSERT INTO FWDemo.ProcessLog (
                ProcessLogMessage,
                CreateDate
        )
        VALUES ('Procedure FWDemo.DailyProcess2 - ERROR',
                GETDATE()
        )
    END

RETURN @RetStat
GO
```

In a "New Query" window in SSMS, execute the code from Listing 1-2 while connected to the FWDemo schema. The code creates two stored procedures that together make up a daily process. With those procedures in place, you can turn your attention to the next problem, which is to schedule those procedures to actually run each day.

Executing the Daily Process

Now that an environment has been built and a process created, let's turn to the execution. Operations staff will have to set up or schedule the procedures to run and either monitor for any error conditions that are raised or set up precedence rules if such a function exists in any scheduling tool used. In a basic sense, we now have a Daily Process that is ready for production. Listing 1-3 shows the statements that can be used to execute the Daily Process procedures and also a SELECT statement that can be run to view the output written to FWDemo.ProcessLog. You will notice that we are ordering the output in a descending order. This will show the most recent messages at the top and eliminate the need to scroll down to get to the messages for the current execution and much easier once the log starts to become heavily populated.

Listing 1-3. Daily Process execute statements and process log SELECT statement

```
EXECUTE FWDemo.DailyProcess1

EXECUTE FWDemo.DailyProcess2

SELECT ProcessLogID
      ,ProcessLogMessage
      ,CreateDate
FROM FWDemo.ProcessLog
ORDER BY ProcessLogID desc
```

Including a Monthly Process

Now it's time to add another layer to our production process. In Listing 1-4, there is code to create two more stored procedures that will make up a Monthly Process. The procedures operate the same as our daily process procedures, and there are some business rules associated with them. First, the monthly process will run on the first day of the month. Second, it will run after the successful execution of the Daily Process, and third, FWDemo.MonthlyProcess1 must complete successfully before FWDemo. MonthlyProcess2 can be executed.

Listing 1-4. Monthly Process stored procedures

```
If Exists(Select s.name + '.' + p.name
          From sys.procedures p
          Join sys.schemas s
            On s.schema_id = p.schema_id
          Where s.name = 'FWDemo'
            And p.name = 'MonthlyProcess1')
begin
  print ' - Dropping FWDemo.MonthlyProcess1 stored procedure'
  Drop Procedure FWDemo.MonthlyProcess1
  print ' - FWDemo.MonthlyProcess1 stored procedure dropped'
end
GO

CREATE PROCEDURE FWDemo.MonthlyProcess1
AS
-----------------------------------------------------------------------------
-----------------------------------------------------------------------------
--
-- Purpose: This procedure is part of the Stored Procedure Framework Demo.
--
-- NOTE: An Error situation can be created for testing/demo purposes by
--     un-commenting the Error code in the body of the procedure.  To return
--     to a procedure with a successful execution, re-comment the code or
--     recompile the original.
--
-----------------------------------------------------------------------------
-----------------------------------------------------------------------------

SET NOCOUNT ON

/*********************************************/
/*  Log the START of the procedure to the process log  */
/***************************** ***********/
```

```
INSERT INTO FWDemo.ProcessLog (
      ProcessLogMessage,
      CreateDate
)
Values ('Procedure FWDemo.MonthlyProcess1 - STARTING',
      GETDATE()
)

DECLARE @RetStat int
SET @RetStat = 0

/*****************************************/
/*  Force an ERROR CONDITION for this procedure  */
/*****************************************/

--INSERT INTO FWDemo.ProcessLog (
--      ProcessLogMessage,
--      CreateDate
--)
--VALUES ('Procedure FWDemo.MonthlyProcess1 - Problem Encountered',
--      GETDATE()
--)
--SET @RetStat = 1

/***************************************************/
/*  Log the COMPLETION of the procedure to the process log    */
/***************************************************/

IF @RetStat = 0
   BEGIN
      INSERT INTO FWDemo.ProcessLog (
            ProcessLogMessage,
            CreateDate
      )
      VALUES ('Procedure FWDemo.MonthlyProcess1 - COMPLETED',
            GETDATE()
      )
   END
```

```
ELSE
   BEGIN
      INSERT INTO FWDemo.ProcessLog (
           ProcessLogMessage,
           CreateDate
      )
      VALUES ('Procedure FWDemo.MonthlyProcess1 - ERROR',
           GETDATE()
      )
   END

RETURN @RetStat
GO

If Exists(Select s.name + '.' + p.name
          From sys.procedures p
          Join sys.schemas s
            On s.schema_id = p.schema_id
          Where s.name = 'FWDemo'
            And p.name = 'MonthlyProcess2')
begin
  print ' - Dropping FWDemo.MonthlyProcess2 stored procedure'
  Drop Procedure FWDemo.MonthlyProcess2
  print ' - FWDemo.MonthlyProcess2 stored procedure dropped'
end
GO

CREATE PROCEDURE FWDemo.MonthlyProcess2
AS
--------------------------------------------------------------------------------
--------------------------------------------------------------------------------
--
-- Purpose: This procedure is part of the Stored Procedure Framework Demo.
--
-- NOTE: An Error situation can be created for testing/demo purposes by
--     un-commenting the Error code in the body of the procedure.  To return
```

```
--      to a procedure with a successful execution, re-comment the code or
--      recompile the original.
--
-------------------------------------------------------------------------
-------------------------------------------------------------------------

SET NOCOUNT ON

/******************************************/
/*  Log the START of the procedure to the process log  */
/******************************************/

INSERT INTO FWDemo.ProcessLog (
      ProcessLogMessage,
      CreateDate
)
Values ('Procedure FWDemo.MonthlyProcess2 - STARTING',
      GETDATE()
)

DECLARE @RetStat int

SET @RetStat = 0

/******************************************/
/*  Force an ERROR CONDITION for this procedure  */
/******************************************/

--INSERT INTO FWDemo.ProcessLog (
--      ProcessLogMessage,
--      CreateDate
--)
--VALUES ('Procedure FWDemo.MonthlyProcess2 - Problem Encountered',
--      GETDATE()
--)
--SET @RetStat = 1

/*********************************************/
```

```
/*  Log the COMPLETION of the procedure to the process log       */
/******************************************************/

IF @RetStat = 0
    BEGIN
        INSERT INTO FWDemo.ProcessLog (
                ProcessLogMessage,
                CreateDate
        )
        VALUES ('Procedure FWDemo.MonthlyProcess2 - COMPLETED',
                GETDATE()
        )
    END
ELSE
    BEGIN
        INSERT INTO FWDemo.ProcessLog (
                ProcessLogMessage,
                CreateDate
        )
        VALUES ('Procedure FWDemo.MonthlyProcess2 - ERROR',
                GETDATE()
        )
    END

RETURN @RetStat
GO
```

The execute statements to be run or scheduled by the operations staff are shown in Listing 1-5. Use the SELECT statement introduced earlier to monitor the progress of the process execution.

Listing 1-5. Monthly Process execute statements

```
EXECUTE FWDemo.MonthlyProcess1

EXECUTE FWDemo.MonthlyProcess2
```

From an operations perspective, we have now introduced more complexity. Process set up/scheduling now has additional layers of precedence and also a timing factor. Scheduling and monitoring have become more critical, and there are more places that require attention in order for everything to operate smoothly.

Summary

A system has been built that consists of two daily and two monthly processes. Dependencies exist that require the successful execution of all prior processes before initiating the execution of the next process. These dependencies must be applied in the scheduling of the processes in the course of regular production and also in troubleshooting and problem remediation. All of these "moving parts" not only add complexity but also points of vulnerability to the system due to the need for outside attention.

CHAPTER 2

Automation with Stored Procedures

Now that we see all the pieces that need to be coordinated day in and day out for executing the processes defined in Chapter 1, let's introduce the idea of controllers. A controller is the most basic element in our demonstration of the framework concept. The examples that follow demonstrate how controllers can make your job easier while also reducing the overhead that your operations staff must deal with.

A Daily Process Controller

For our purposes, a controller is a stored procedure that itself executes all the elements of a process and can address business rules within the process. Listing 2-1 shows the code to create a stored procedure called FWDemo.DailyProcessController. This stored procedure executes both of our Daily Process stored procedures. It logs its own progress and checks the return status of each of the stored procedures to ensure successful completion of each.

Listing 2-1. Daily Process Controller stored procedure

```
If Exists(Select s.name + '.' + p.name
        From sys.procedures p
        Join sys.schemas s
          On s.schema_id = p.schema_id
        Where s.name = 'FWDemo'
          And p.name = 'DailyProcessController')
begin
```

© Andy Leonard, Kent Bradshaw 2020
A. Leonard and K. Bradshaw, *SQL Server Data Automation Through Frameworks*,
https://doi.org/10.1007/978-1-4842-6213-9_2

```
  print ' - Dropping FWDemo.DailyProcessController stored procedure'
  Drop Procedure FWDemo.DailyProcessController
  print ' - FWDemo.DailyProcessController stored procedure dropped'
end
GO

CREATE PROCEDURE FWDemo.DailyProcessController
AS
----------------------------------------------------------------------------
----------------------------------------------------------------------------
--
-- Purpose: This procedure is part of the Stored Procedure Framework Demo.
--          It is the Controller for the Daily Process Stored Procedures.
--
----------------------------------------------------------------------------
----------------------------------------------------------------------------

SET NOCOUNT ON

/*********************************************/
/*  Log the START of the procedure to the process log  */
/*********************************************/

INSERT INTO FWDemo.ProcessLog (
      ProcessLogMessage,
      CreateDate
)
Values ('Procedure FWDemo.DailyProcessController - STARTING',
      GETDATE()
)

DECLARE @RetStat int

SET @RetStat = 0

/***************************************/
/*  Execute the DailyProcess1 Stored Procedures  */
/***************************************/
```

```
EXEC @RetStat = FWDemo.DailyProcess1

IF @RetStat <> 0
      GOTO EndController

/****************************************/
/*  Execute the DailyProcess2 Stored Procedures   */
/****************************************/

EXEC @RetStat = FWDemo.DailyProcess2

IF @RetStat <> 0
      GOTO EndController

/****************************************************/
/*  Log the COMPLETION of the procedure to the process log     */
/****************************************************/
EndController:

IF @RetStat = 0
   BEGIN
      INSERT INTO FWDemo.ProcessLog (
            ProcessLogMessage,
            CreateDate
      )
      VALUES ('Procedure FWDemo.DailyProcessController - COMPLETED',
            GETDATE()
      )
   END
ELSE
   BEGIN
      INSERT INTO FWDemo.ProcessLog (
            ProcessLogMessage,
            CreateDate
      )
```

```
    VALUES ('Procedure FWDemo.DailyProcessController - ERROR',
        GETDATE()
    )
  END

RETURN @RetStat
GO
```

With the controller in Listing 2-1, execution by the operations staff has just been made simpler. Instead of scheduling both the stored procedures and dealing with precedence rules from the schedule, only the controller execution needs to be set up, and monitoring for successful completion is now easier. Listing 2-2 shows the code used to execute the Daily Process controller. As always, FWDemo.ProcessLog shows the progress of the execution.

Listing 2-2. Daily Process Controller execute statement

```
EXECUTE FWDemo.DailyProcessController
```

Another advantage of using the controller pattern is the situation where a new stored procedure and/or new business rules need to be added to a process. Without the controller, the new stored procedure is compiled, and it is the responsibility of the operations staff to execute it in its proper position in the process and also account for the new business rule requirements in the setup process. That is a lot of steps that need to happen after development/testing has ended and, thus, an opportunity for error. With a controller in place, development and testing would include not only the new stored procedure(s) but also the necessary modifications to the controller stored procedure to execute the new stored procedures. The business rules then become part of the development/testing effort, and implementation to production becomes as simple as compiling the new stored procedure(s) and the new version of the controller stored procedure. Since the controller execution is already scheduled, no other changes are necessary for the operations staff. We have eliminated a lot of the manual intervention when deploying to production and, therefore, reduced opportunities for errors.

Monthly Process Controller

Next, we will add a controller for our Monthly Process. Listing 2-3 has the code for the stored procedure called FWDemo.MonthlyProcessController. This stored procedure executes both of the monthly process stored procedures and handles the precedence rule within the process while logging its progress in FWDemo.ProgressLog. The operations staff can simply execute this controller to start the process.

Listing 2-3. Monthly Process Controller stored procedure

```
If Exists(Select s.name + '.' + p.name
        From sys.procedures p
        Join sys.schemas s
          On s.schema_id = p.schema_id
        Where s.name = 'FWDemo'
          And p.name = 'MonthlyProcessController')
begin
  print ' - Dropping FWDemo.MonthlyProcessController stored procedure'
  Drop Procedure FWDemo.MonthlyProcessController
  print ' - FWDemo.MonthlyProcessController stored procedure dropped'
end
GO

CREATE PROCEDURE FWDemo.MonthlyProcessController
AS
----------------------------------------------------------------------------
----------------------------------------------------------------------------
--
-- Purpose: This procedure is part of the Stored Procedure Framework Demo.
--          It is the Controller for the Monthly Process Stored Procedures.
--
----------------------------------------------------------------------------
----------------------------------------------------------------------------
```

```
SET NOCOUNT ON

/*********************************************/
/*  Log the START of the procedure to the process log   */
/*********************************************/

INSERT INTO FWDemo.ProcessLog (
      ProcessLogMessage,
      CreateDate
)
Values ('Procedure FWDemo.MonthlyProcessController - STARTING',
      GETDATE()
)

DECLARE @RetStat int

SET @RetStat = 0

/*********************************************/
/*  Execute the MonthlyProcess1 Stored Procedures   */
/*********************************************/

EXEC @RetStat = FWDemo.MonthlyProcess1

IF @RetStat <> 0
      GOTO EndController

/*********************************************/
/*  Execute the MonthlyProcess2 Stored Procedures   */
/*********************************************/

EXEC @RetStat = FWDemo.MonthlyProcess2

IF @RetStat <> 0
      GOTO EndController

/*************************************************/
/*  Log the COMPLETION of the procedure to the process log    */
/*************************************************/
```

```
EndController:

IF @RetStat = 0
    BEGIN
        INSERT INTO FWDemo.ProcessLog (
                ProcessLogMessage,
                CreateDate
        )
        VALUES ('Procedure FWDemo.MonthlyProcessController - COMPLETED',
                GETDATE()
        )
    END
ELSE
    BEGIN
        INSERT INTO FWDemo.ProcessLog (
                ProcessLogMessage,
                CreateDate
        )
        VALUES ('Procedure FWDemo.MonthlyProcessController - ERROR',
                GETDATE()
        )
    END

RETURN @RetStat
GO
```

Listing 2-4 has the statement used to execute the new Monthly Process Controller.

Listing 2-4. Monthly Process Controller execute statement

```
EXECUTE FWDemo.MonthlyProcessController
```

Not addressed here, however, are the business precedence rules between controllers that are 1) the Monthly Process executes on the first day of the month and 2) after the successful completion of the Daily Process. But, these details for the operations staff to address have become much simpler with the introduction of the controller pattern.

Summary

Controllers incorporate dependencies, defined by the business rules, and coordinate the execution of processes. There are two primary advantages that come from this. First, it makes scheduling much easier because the dependencies are built in. Second, it provides ready access and reference to the process flow when in the midst of troubleshooting process issues.

CHAPTER 3

Stored Procedure Orchestrators

Building on the controller pattern in Chapter 2, the next logical progression for the framework is the concept of the orchestrator. It is basically a controller for controllers and can address the issue of precedence rules between controllers.

The Orchestrator

Controllers are themselves a type of process just as the daily and monthly processes that have been built. We may find that dependencies exist between the processing of different controllers. That is where an orchestrator comes in. Orchestrators can address those dependencies in exactly the same way as controllers. Listing 3-1 shows the code to create a stored procedure called FWDemo.ProcessOrchestrator. This stored procedure executes the Daily Controller stored procedure and, when appropriate based on the date, the Monthly Controller stored procedure.

Listing 3-1. Process Orchestrator stored procedure

```
IF EXISTS (SELECT s.name + '.' + p.name
        FROM sys.procedures p
        JOIN sys.schemas s
          ON s.schema_id = p.schema_id
        WHERE s.name = 'FWDemo'
          AND p.name = 'ProcessOrchestrator')
```

© Andy Leonard, Kent Bradshaw 2020
A. Leonard and K. Bradshaw, *SQL Server Data Automation Through Frameworks*,
https://doi.org/10.1007/978-1-4842-6213-9_3

```
BEGIN
  PRINT ' - Dropping FWDemo.ProcessOrchestrator stored procedure'
  DROP PROCEDURE FWDemo.ProcessOrchestrator
  PRINT ' - FWDemo.ProcessOrchestrator stored procedure dropped'
END
GO

CREATE PROCEDURE FWDemo.ProcessOrchestrator
      @RunDate smalldatetime = null
AS
-------------------------------------------------------------------------
-------------------------------------------------------------------------
--
-- Purpose: This procedure is part of the Stored Procedure Framework
Demo.                       --     It is the Orchestrator Procedure that
executes the Controller          --     Procedures.
--
-------------------------------------------------------------------------
-------------------------------------------------------------------------

SET NOCOUNT ON
/****************************************************/
/*  Log the START of the procedure to the process log  */
/****************************************************/

INSERT INTO FWDemo.ProcessLog (
      ProcessLogMessage,
      CreateDate
)
VALUES ('Procedure FWDemo.ProcessOrchestrator - STARTING',
      GETDATE()
)

IF @RunDate is null
      SET @RunDate = GETDATE()

DECLARE @RetStat int

SET @RetStat = 0
```

28

```
/******************************************/
/*  Execute the Daily Process Controller  */
/******************************************/

EXEC @RetStat = FWDemo.DailyProcessController

IF @RetStat <> 0
    GOTO EndOrchestrator

/*********************************************************************/
/*  Execute the MonthlyProcessController IF @RunDate is the first day */
/*     of a month                                                  */
/*********************************************************************/

IF DATEPART(DAY, @RunDate) = 1
   BEGIN
      EXEC @RetStat = FWDemo.MonthlyProcessController

      IF @RetStat <> 0
           GOTO EndOrchestrator
   END

/****************************************************************/
/*  Log the COMPLETION of the procedure to the process log   */
/****************************************************************/

EndOrchestrator:

IF @RetStat = 0
   BEGIN
      INSERT INTO FWDemo.ProcessLog (
           ProcessLogMessage,
           CreateDate
      )
      VALUES ('Procedure FWDemo.ProcessOrchestrator - COMPLETED',
           GETDATE()
      )
   END
```

```
ELSE
    BEGIN
        INSERT INTO FWDemo.ProcessLog (
                ProcessLogMessage,
                CreateDate
        )
        VALUES ('Procedure FWDemo.ProcessOrchestrator - ERROR',
                GETDATE()
        )
    END

RETURN @RetStat
GO
```

As with the various stored procedures we have already seen, the orchestrator shows execution progress by logging to FWDemo.ProcessLog. The return status is used to check for successful completion of the controllers. Also, there is code included to determine if it is the first day of the month which is one of the precedence rules for the execution of the Monthly Process. There is an input parameter called "@RunDate" which has a default value of "null." The null value caused the stored procedure to use the current date for processing. However, to make it easy to experiment with the demonstration system, a date can be supplied at execution time to establish the value of "@RunDate". Listing 3-2 shows the execute statements for the orchestrator for using both the current date and a date override. For all practical purposes, all that the operations staff needs to do is set up/schedule the execution of FWDemo.ProcessOrchestrator, and everything else is handled within the patterns that have been built.

Listing 3-2. Process Orchestrator execute statements

```
--IF rundate is current date:
EXECUTE FWDemo.ProcessOrchestrator

--IF rundate is not current date:
EXECUTE FWDemo.ProcessOrchestrator
        @RunDate = '<date>'

--NOTE: <date> is in the format of CCYYMMDD
```

A Quick Process Overview

At this point, let's review what has been built up until now. There is a Daily Process comprised of two stored procedures and a Monthly Process also comprised of two stored procedures. The daily process controller, FWDemo.DailyProcessController, executes the daily process and ensures that the stored procedure FWDemo.DailyProcess1 completes successfully before executing the FWDemo.DailyProcess2 stored procedure. The monthly process controller, FWDemo.MonthlyProcessController, executes the monthly process and ensures that the stored procedure FWDemo.MonthlyProcess1 completes successfully before executing the FWDemo.MontlyProcess2 stored procedure. Wrapped around it all is FWDemo.ProcessOrchestrator that executes the daily process controller and, upon successful completion, checks for a run date that is the first day of the month and, if true, executes the monthly process controller. A single execute statement drives the entire process.

A Look at Troubleshooting Issues

Up until now, our attention has been focused on following the successful completion of the processes that have been built. So, it's time for a reality check. How do we handle error situations? What do we do if one of the processes in the sequence has an error and doesn't complete successfully? That can sometimes be accompanied by a phone call in the middle of the night and some action needs to be taken. Diving in and troubleshooting the problem begins. Looking at FWDemo.ProcessLog is a great place to start in order to identify the location and cause of the problem in order for correction to begin. Once the solution is determined and the problem corrected, how do we complete the remaining steps in the sequence? With our current patterns in place, about the only option is to deconstruct the process where the error occurred and finish manually. After that, we find an execution at the next highest level of our patterns and manually execute from there.

For example, the production process has a problem with the stored procedure FWDemo.DailyProcess2. We research and correct the issue. At this point, FWDemo.DailyProcess2 needs to be manually executed. Upon successful completion, we know that everything in FWDemo.DailyProcessController has finished. The next highest level is the FWDemo.ProcessOrchestrator stored procedure. After the execution of the daily process controller, the orchestrator checks to see if the value in the "@RunDate" parameter equals the first of the month. If that is the case, then we would need to manually execute FWDemo.MonthlyProcessController. If it is not the first of the month, nothing else is necessary. In our system, it is pretty simple. But, in reality, the resolution could have a lot more moving parts and a lot of manual intervention. The more complex the process becomes, the more complex the problem resolution becomes.

Summary

Orchestrators are simply a higher level of controller. It manages the execution of multiple controllers while incorporating any dependencies that may exist between the controllers. It can further simplify scheduling while providing a more complete picture of process requirements that can aid in successful problem remediation.

CHAPTER 4

A Stored Procedure–Based Metadata-Driven Framework

In order to add flexibility and control over the execution of all of the processes within the framework, metadata can be added to the environment. It can identify the stored procedures, controllers, and orchestrators as well as the relationships between each. The metadata will be used as the communication that drives the execution of the processes.

Building the Metadata

If we're going to create metadata, then we need a place in which to store it. We can create a table or a set of tables for that purpose. The metadata for this chapter's example is stored in a set of tables used to control executions. The first table is FWDemo. ProcessProcedure, seen in Listing 4-1.

Listing 4-1. ProcessProcedure table creation

```
IF EXISTS (SELECT * FROM sys.objects
          WHERE object_id = OBJECT_ID(N'FWDemo.ProcessProcedure')
          AND type in (N'U'))
    DROP TABLE FWDemo.ProcessProcedure
GO

SET ANSI_NULLS ON
GO
```

© Andy Leonard, Kent Bradshaw 2020

A. Leonard and K. Bradshaw, *SQL Server Data Automation Through Frameworks*,
https://doi.org/10.1007/978-1-4842-6213-9_4

```
SET QUOTED_IDENTIFIER ON
GO

SET ANSI_PADDING ON
GO

CREATE TABLE [FWDemo].[ProcessProcedure](
    [ProcessProcedureID]    [int] IDENTITY(1,1) NOT NULL,
    [ProcessProcedureName]  [nvarchar](255)      NOT NULL
)
GO

SET ANSI_PADDING OFF
GO
```

This table identifies all of the stored procedures available for execution in the framework. The stored procedures become the building blocks that are collected together in what we will call framework applications. To establish the application, we use the FWDemo.Application table seen in Listing 4-2.

Listing 4-2. Application table creation

```
IF EXISTS (SELECT * FROM sys.objects
            WHERE object_id = OBJECT_ID(N'FWDemo.Application')
             AND type in (N'U'))
    DROP TABLE FWDemo.Application
GO

SET ANSI_NULLS ON
GO

SET QUOTED_IDENTIFIER ON
GO

SET ANSI_PADDING ON
GO
```

```
CREATE TABLE [FWDemo].[Application](
    [ApplicationID]   [int] IDENTITY(1,1) NOT NULL,
    [ApplicationName] [nvarchar](255)     NOT NULL
)
GO

SET ANSI_PADDING OFF
GO
```

Now it is necessary to identify which of the stored procedures are going to be included in each of the defined applications. For that, we use the table, seen in Listing 4-3, called FWDemo.ApplicationProcessProcedure. This table contains the metadata that will define the execution of the process. As we discussed in Chapter 1, building small unit of work stored procedures allows for code reuse. Using the metadata tables allows relating a single stored procedure to multiple applications. A data archive stored procedure would be a good example of where this might be helpful.

Listing 4-3. ApplicationProcessProcedure table creation

```
IF EXISTS (SELECT * FROM sys.objects
    WHERE object_id = OBJECT_ID(N'FWDemo.ApplicationProcessProcedure')
          AND type in (N'U'))
    DROP TABLE FWDemo.ApplicationProcessProcedure
GO

SET ANSI_NULLS ON
GO

SET QUOTED_IDENTIFIER ON
GO

SET ANSI_PADDING ON
GO

CREATE TABLE [FWDemo].[ApplicationProcessProcedure](
    [ApplicationProcessProcedureID] [int] IDENTITY(1,1) NOT NULL,
    [ApplicationID]          [int]  NOT NULL,
    [ProcessProcedureID]     [int]  NOT NULL,
```

```
        [ExecutionOrder]           [int]  NOT NULL,
        [Active]                   [bit]  NOT NULL
)
GO

SET ANSI_PADDING OFF
GO
```

There is a column named ExecutionOrder which will determine where in the process sequence its associated stored procedure will be executed. Also, the Active column indicates that a stored procedure is enabled for execution. It is used to disable procedures that, for some reason, are not to be executed in the process. This is great if you want to temporarily bypass a procedure and will be discussed later in this chapter.

With the metadata tables created, let's load the data for our processes. Listing 4-4 shows the code that loads all of our stored procedures into the FWDemo. ProcessProcedure table. Only the name is inserted because the value for ProcessProcedureID, being an IDENTITY column, will be automatically generated.

Listing 4-4. Load data to the ProcessProcedure table

```
--Load First Daily Procedure
INSERT INTO FWDemo.ProcessProcedure (ProcessProcedureName)
VALUES ('FWDemo.DailyProcess1')
GO

--Load Second Daily Procedure
INSERT INTO FWDemo.ProcessProcedure (ProcessProcedureName)
VALUES ('FWDemo.DailyProcess2')
GO

--Load First Monthly Procedure
INSERT INTO FWDemo.ProcessProcedure (ProcessProcedureName)
VALUES ('FWDemo.MonthlyProcess1')
GO

--Load Second Monthly Procedure
INSERT INTO FWDemo.ProcessProcedure (ProcessProcedureName)
VALUES ('FWDemo.MonthlyProcess2')
GO
```

Listing 4-5 has the code to define the applications for our demonstration into the table FWDemo.Application. For our basic framework, we are defining our daily and monthly processes. Again, the value of the IDENTITY column ApplicationID will be automatically generated.

Listing 4-5. Load data to the Application table

```
--Load Application for Daily Process
INSERT INTO FWDemo.Application (ApplicationName)
VALUES ('DailyProcessControllerMD')
GO
--Load Application for Monthly Process
INSERT INTO FWDemo.Application (ApplicationName)
VALUES ('MonthlyProcessControllerMD')
GO
```

To put all of the pieces together, Listing 4-6 contains the code to relate the stored procedures to their intended application. Notice the values used for the execution order. Using values in a scale, in this case a scale of ten, leaves gaps between the initially defined execution order. This way, if a new procedure is added later on that needs to execute between existing stored procedures, there is room to add a value, and the existing execution order values do not have to be renumbered. Also, notice that a value of "1" is used for the Active column. This enables the procedure for execution within the application. A value of "0" will disable the procedure in the context of the application.

Listing 4-6. Load metadata for framework

```
DECLARE @AppID     int,
        @ProcID    int

/******************************/
/*  Daily Process Application   */
/******************************/

SET @AppID = (SELECT ApplicationID
              FROM FWDemo.Application
              WHERE ApplicationName = 'DailyProcessControllerMD')
```

```
SET @ProcID = (SELECT ProcessProcedureID
            FROM FWDemo.ProcessProcedure
            WHERE ProcessProcedureName = 'FWDemo.DailyProcess1')

INSERT INTO FWDemo.ApplicationProcessProcedure (
      ApplicationID,
      ProcessProcedureID,
      ExecutionOrder,
      Active
)
VALUES (@AppID,
      @ProcID,
      10,
      1
)

SET @ProcID = (SELECT ProcessProcedureID
            FROM FWDemo.ProcessProcedure
            WHERE ProcessProcedureName = 'FWDemo.DailyProcess2')

INSERT INTO FWDemo.ApplicationProcessProcedure (
      ApplicationID,
      ProcessProcedureID,
      ExecutionOrder,
      Active
)
VALUES (@AppID,
      @ProcID,
      20,
      1
)

/********************************/
/*  Monthly Process Application   */
/********************************/

SET @AppID = (SELECT ApplicationID
            FROM FWDemo.Application
            WHERE ApplicationName = 'MonthlyProcessControllerMD')
```

```
SET @ProcID = (SELECT ProcessProcedureID
          FROM FWDemo.ProcessProcedure
          WHERE ProcessProcedureName = 'FWDemo.MonthlyProcess1')
INSERT INTO FWDemo.ApplicationProcessProcedure (
      ApplicationID,
      ProcessProcedureID,
      ExecutionOrder,
      Active
)
VALUES (@AppID,
      @ProcID,
      10,
      1
)

SET @ProcID = (SELECT ProcessProcedureID
          FROM FWDemo.ProcessProcedure
          WHERE ProcessProcedureName = 'FWDemo.MonthlyProcess2')

INSERT INTO FWDemo.ApplicationProcessProcedure (
      ApplicationID,
      ProcessProcedureID,
      ExecutionOrder,
      Active
)
VALUES (@AppID,
      @ProcID,
      20,
      1
)
```

Metadata-Ready Controllers

All of the metadata is now in place. But, in order to execute using the metadata, new controllers are needed to use the metadata. Listing 4-7 has the code for the new daily and monthly controllers. You will notice that the names of these new procedures are suffixed with "MD".

Listing 4-7. New metadata controllers

```
IF EXISTS (SELECT s.name + '.' + p.name
          FROM sys.procedures p
          JOIN sys.schemas s
            ON s.schema_id = p.schema_id
          WHERE s.name = 'FWDemo'
            AND p.name = 'DailyProcessControllerMD')
BEGIN
  PRINT ' - Dropping FWDemo.DailyProcessControllerMD stored procedure'
  DROP PROCEDURE FWDemo.DailyProcessControllerMD
  PRINT ' - FWDemo.DailyProcessControllerMD stored procedure dropped'
END
GO

CREATE PROCEDURE FWDemo.DailyProcessControllerMD
AS
--------------------------------------------------------------------------------
--------------------------------------------------------------------------------
--
-- Purpose: This procedure is part of the Stored Procedure Framework
Demo.                    --    It is the Metadata Controller version for
the Daily Process Stored    --    Procedures.
--
--------------------------------------------------------------------------------
--------------------------------------------------------------------------------

SET NOCOUNT ON
```

```
/*******************************************************/
/*  Log the START of the procedure to the process log  */
/*******************************************************/

INSERT INTO FWDemo.ProcessLog (
     ProcessLogMessage,
     CreateDate
)
VALUES ('Procedure FWDemo.DailyProcessControllerMD - STARTING',
     GETDATE()
)

DECLARE @RetStat int

SET @RetStat = 0

/***************************************************************/
/*  Get and Execute the Active DailyProcess Stored Procedures  */
/***************************************************************/

DECLARE @DailyProcName  nvarchar(255)
     ,@AppID     int
     ,@AppPPID   int

DECLARE @DailyCursor as CURSOR

SET @DailyCursor = CURSOR FOR
SELECT pp.ProcessProcedureName, a.ApplicationID,
     app.ApplicationProcessProcedureID
FROM FWDemo.Application a
JOIN FWDemo.ApplicationProcessProcedure app
     ON a.ApplicationID = app.ApplicationID
JOIN FWDemo.ProcessProcedure pp
     ON app.ProcessProcedureID = pp.ProcessProcedureID
WHERE a.ApplicationName = 'DailyProcessControllerMD'
  AND app.Active = 1
ORDER BY app.ExecutionOrder

OPEN @DailyCursor
FETCH NEXT FROM @DailyCursor INTO @DailyProcName, @AppID, @AppPPID
```

```
WHILE @@FETCH_STATUS = 0
    BEGIN
        INSERT INTO FWDemo.ProcessLog (
                ProcessLogMessage,
                CreateDate
        )
        VALUES ('Executing ' + @DailyProcName + ' ApplicationID = ' +
                convert(varchar(10), @AppID) +
                ' ApplicationProcessProcedureID = ' + convert(varchar(10),
                @AppPPID),
                GETDATE()
        )

        EXEC @RetStat = @DailyProcName

        IF @RetStat <> 0
            BEGIN
                CLOSE @DailyCursor
                DEALLOCATE @DailyCursor
                GOTO EndController
            END

        FETCH NEXT FROM @DailyCursor INTO @DailyProcName, @AppID, @AppPPID
    END

CLOSE @DailyCursor
DEALLOCATE @DailyCursor

/***********************************************************/
/*  Log the COMPLETION of the procedure to the process log  */
/***********************************************************/
EndController:

IF @RetStat = 0
    BEGIN
        INSERT INTO FWDemo.ProcessLog (
                ProcessLogMessage,
                CreateDate
        )
```

```
        VALUES ('Procedure FWDemo.DailyProcessControllerMD - COMPLETED',
            GETDATE()
        )
    END
ELSE
    BEGIN
        INSERT INTO FWDemo.ProcessLog (
            ProcessLogMessage,
            CreateDate
        )
        VALUES ('Procedure FWDemo.DailyProcessControllerMD - ERROR',
            GETDATE()
        )
    END

RETURN @RetStat
GO

IF EXISTS (SELECT s.name + '.' + p.name
            FROM sys.procedures p
            JOIN sys.schemas s
              ON s.schema_id = p.schema_id
            WHERE s.name = 'FWDemo'
              AND p.name = 'MonthlyProcessControllerMD')
BEGIN
  PRINT ' - Dropping FWDemo.MonthlyProcessControllerMD stored procedure'
  DROP Procedure FWDemo.MonthlyProcessControllerMD
  PRINT ' - FWDemo.MonthlyProcessControllerMD stored procedure dropped'
END
GO

CREATE PROCEDURE FWDemo.MonthlyProcessControllerMD
AS
--------------------------------------------------------------------------------
--------------------------------------------------------------------------------
```

```
--
-- Purpose: This procedure is part of the Stored Procedure Framework Demo.
--     It is the Metadata Controller version for the Monthly Process Stored
--     Procedures.
--
---------------------------------------------------------------------------
---------------------------------------------------------------------------

SET NOCOUNT ON

/*****************************************************/
/*  Log the START of the procedure to the process log  */
/*****************************************************/

INSERT INTO FWDemo.ProcessLog (
     ProcessLogMessage,
     CreateDate
)
VALUES ('Procedure FWDemo.MonthlyProcessControllerMD - STARTING',
     GETDATE()
)

DECLARE @RetStat int

SET @RetStat = 0

/***********************************************************************/
/*  Get and Execute the Active MonthlyProcess Stored Procedures  */
/***********************************************************************/

DECLARE @MonthlyProcName nvarchar(255)
     ,@AppID     int
     ,@AppPPID   int

DECLARE @MonthlyCursor as CURSOR

SET @MonthlyCursor = CURSOR FOR
SELECT pp.ProcessProcedureName, a.ApplicationID,
     app.ApplicationProcessProcedureID
FROM FWDemo.Application a
```

```
JOIN FWDemo.ApplicationProcessProcedure app
     ON a.ApplicationID = app.ApplicationID
JOIN FWDemo.ProcessProcedure pp
     ON app.ProcessProcedureID = pp.ProcessProcedureID
WHERE a.ApplicationName = 'MonthlyProcessControllerMD'
  AND app.Active = 1
ORDER BY app.ExecutionOrder

OPEN @MonthlyCursor
FETCH NEXT FROM @MonthlyCursor INTO @MonthlyProcName, @AppID, @AppPPID

WHILE @@FETCH_STATUS = 0
   BEGIN
      INSERT INTO FWDemo.ProcessLog (
            ProcessLogMessage,
            CreateDate
      )
      VALUES (
            'Executing ' + @DailyProcName + ' ApplicationID = ' +
            convert(varchar(10), @AppID) +
            ' ApplicationProcessProcedureID = ' + convert(varchar(10),
            @AppPPID),
            GETDATE()
      )

      EXEC @RetStat = @MonthlyProcName

      IF @RetStat <> 0
         BEGIN
            CLOSE @MonthlyCursor
            DEALLOCATE @MonthlyCursor
            GOTO EndController
         END

      FETCH NEXT FROM @MonthlyCursor INTO @MonthlyProcName, @AppID,
      @AppPPID
   END
```

```
CLOSE @MonthlyCursor
DEALLOCATE @MonthlyCursor

/***************************************************************/
/*  Log the COMPLETION of the procedure to the process log    */
/***************************************************************/
EndController:

IF @RetStat = 0
    BEGIN
        INSERT INTO FWDemo.ProcessLog (
                ProcessLogMessage,
                CreateDate
        )
        VALUES ('Procedure FWDemo.MonthlyProcessControllerMD - COMPLETED',
                GETDATE()
        )
    END
ELSE
    BEGIN
        INSERT INTO FWDemo.ProcessLog (
                ProcessLogMessage,
                CreateDate
        )
        VALUES ('Procedure FWDemo.MonthlyProcessControllerMD - ERROR',
                GETDATE()
        )
    END

RETURN @RetStat
GO
```

As you can see, these new versions of the controllers use a cursor to retrieve all of the active stored procedures related to the respective applications, sorted by the execution order. The procedure names, along with the values of all of the associated keys, are loaded into variables. The stored procedure name is used to direct the execution while the key values are used in the messages written to FWDemo.ProcessLog to track execution and aid in monitoring or troubleshooting. The cursor loops through each stored procedure until all have been executed or an error occurs.

Metadata-Ready Orchestrator

With the new versions of the controllers in place, a new version of the orchestrator, also with the name suffixed with "MD", is needed to execute them. Listing 4-8 has the code for the new orchestrator. Note, separate versions of the controllers and orchestrator were created to keep the original versions in place and operational so that all aspects of the framework demonstration remain intact and functional.

Listing 4-8. New metadata orchestrator

```
IF EXISTS (SELECT s.name + '.' + p.name
          FROM sys.procedures p
          JOIN sys.schemas s
            ON s.schema_id = p.schema_id
          WHERE s.name = 'FWDemo'
            AND p.name = 'ProcessOrchestratorMD')
BEGIN
  PRINT ' - Dropping FWDemo.ProcessOrchestratorMD stored procedure'
  DROP PROCEDURE FWDemo.ProcessOrchestratorMD
  PRINT ' - FWDemo.ProcessOrchestratorMD stored procedure dropped'
END
GO

CREATE PROCEDURE FWDemo.ProcessOrchestratorMD
    @RunDate smalldatetime = null
AS
-------------------------------------------------------------------------
-------------------------------------------------------------------------
--
```

```
-- Purpose: This procedure is part of the Stored Procedure Framework Demo.
--    It is the Orchestrator Procedure that executes the Metadate version
--    of the Controller Procedures.
--
--------------------------------------------------------------------------
--------------------------------------------------------------------------

SET NOCOUNT ON

/*********************************************************/
/*  Log the START of the procedure to the process log  */
/*********************************************************/

INSERT INTO FWDemo.ProcessLog (
     ProcessLogMessage,
     CreateDate
)
Values ('Procedure FWDemo.ProcessOrchestratorMD - STARTING',
     GETDATE()
)

IF @RunDate is null
     SET @RunDate = GETDATE()

DECLARE @RetStat int

SET @RetStat = 0

/*******************************************/
/*  Execute the DailyProcessControllerMD   */
/*******************************************/

EXEC @RetStat = FWDemo.DailyProcessControllerMD

IF @RetStat <> 0
     GOTO EndOrchestratorMD

/*****************************************************************************/
/*  Execute the MonthlyProcessControllerMD IF @RunDate is the first day */
/*  of a month                                                         */
/*****************************************************************************/
```

```
IF DATEPART(DAY, @RunDate) = 1
   BEGIN
      EXEC @RetStat = FWDemo.MonthlyProcessControllerMD

      IF @RetStat <> 0
            GOTO EndOrchestratorMD
   END

/**********************************************************/
/*  Log the COMPLETION of the procedure to the process log  */
/**********************************************************/
EndOrchestratorMD:

IF @RetStat = 0
   BEGIN
      INSERT INTO FWDemo.ProcessLog (
            ProcessLogMessage,
            CreateDate
      )
      VALUES ('Procedure FWDemo.ProcessOrchestratorMD - COMPLETED',
            GETDATE()
      )
   END
ELSE
   BEGIN
      INSERT INTO FWDemo.ProcessLog (
            ProcessLogMessage,
            CreateDate
      )
      VALUES ('Procedure FWDemo.ProcessOrchestratorMD - ERROR',
            GETDATE()
      )
   END

RETURN @RetStat
GO
```

Just like the original version, the only thing that needs to be set up/scheduled for execution is the orchestrator with the optional @RunDate parameter override. Listing 4-9 has the execution commands.

Listing 4-9. Metadata version of Process Orchestrator execute statements

```
--IF rundate is current date:
EXECUTE FWDemo.ProcessOrchestratorMD

--IF rundate is not current date:
EXECUTE FWDemo.ProcessOrchestratorMD
    @RunDate = '<date>'

--NOTE: <date> is in the format of CCYYMMDD
```

Troubleshooting with Metadata

This is the most basic metadata setup for the system we have built. So how can executions be completed in an error resolution situation like we ended up with in Chapter 3? Let's say that FWDemo.DailyProcess2 was unsuccessful. Listing 4-10 shows the messages from FWDemo.ProcessLog that shows that the error occurred.

Listing 4-10. Output from the Process Log

```
ProcessLogID      ProcessLogMessage
CreateDate
39                Procedure FWDemo.ProcessOrchestratorMD - ERROR
2018-09-17 20:47:00
38                Procedure FWDemo.DailyProcessControllerMD - ERROR
2018-09-17 20:47:00
37                Procedure FWDemo.DailyProcess2 - Problem Encountered
2018-09-17 20:47:00
36                Procedure FWDemo.DailyProcess2 - STARTING
2018-09-17 20:46:00
35                Executing FWDemo.DailyProcess2 ApplicatonID = 1
2018-09-17 20:46:00
                  ApplicatonProcessProcedureID = 2
```

```
34                    Procedure FWDemo.DailyProcess1 - COMPLETED
2018-09-17 20:46:00
33                    Procedure FWDemo.DailyProcess1 - STARTING
2018-09-17 20:45:00
32                    Executing FWDemo.DailyProcess1 ApplicatonID = 1
2018-09-17 20:45:00
                      ApplicatonProcessProcedureID = 1
31                    Procedure FWDemo.DailyProcessControllerMD - STARTING
2018-09-17 20:45:00                        .
30                         Procedure FWDemo.ProcessOrchestratorMD -
STARTING                   2018-09-17 20:45:00
```

Once the resolution is determined, we can see from the messages that FWDemo.
DailyProcess1 has already run successfully and therefore does not have to be rerun. We
see in the message that the ApplicationProcessProcedureID value is 1. We can use the
code in Listing 4-11 to disable the FWDemo.DailyProcess1 so that it will not run if the
application is executed.

Listing 4-11. Code to disable execution

```
--DISABLE stored procedure from execution for RESTART

UPDATE FWDemo.ApplicationProcessProcedure
    SET Active = 0
WHERE ApplicationProcessProcedureID = 1     -- or value from Process Log
```

At this point, the scheduled job can be restarted as before, and the execution will
start with FWDemo.DailyProcess2 as shown in Listing 4-12. Once the job is completed,
simply reset the Active column for FWDemo.DailyProcess1 with code like that found in
Listing 4-13.

Listing 4-12. Process Log after restart

```
ProcessLogID      ProcessLogMessage
CreateDate
54                    Procedure FWDemo.ProcessOrchestratorMD - COMPLETED
2018-09-17 20:58:00
```

51

```
53                Procedure FWDemo.MonthlyProcessControllerMD - COMPLETED
2018-09-17 20:58:00
52                Procedure FWDemo.MonthlyProcess2 - COMPLETED
2018-09-17 20:58:00
51                Procedure FWDemo.MonthlyProcess2 - STARTING
2018-09-17 20:57:00
50                Executing FWDemo.MonthlyProcess2 ApplicatonID = 2
2018-09-17 20:57:00
                  ApplicatonProcessProcedureID = 4
49                Procedure FWDemo.MonthlyProcess1 - COMPLETED
2018-09-17 20:57:00
48                Procedure FWDemo.MonthlyProcess1 - STARTING
2018-09-17 20:56:00
47                Executing FWDemo.MonthlyProcess1 ApplicatonID = 2
2018-09-17 20:56:00
                  ApplicatonProcessProcedureID = 3
46                Procedure FWDemo.MonthlyProcessControllerMD - STARTING
2018-09-17 20:56:00
45                Procedure FWDemo.DailyProcessControllerMD - COMPLETED
2018-09-17 20:56:00
44                Procedure FWDemo.DailyProcess2 - COMPLETED
2018-09-17 20:56:00
43                Procedure FWDemo.DailyProcess2 - STARTING
2018-09-17 20:56:00
42                Executing FWDemo.DailyProcess2 ApplicatonID = 1
2018-09-17 20:55:00
                  ApplicatonProcessProcedureID = 2
41                Procedure FWDemo.DailyProcessControllerMD - STARTING
2018-09-17 20:55:00
40                Procedure FWDemo.ProcessOrchestratorMD - STARTING
2018-09-17 20:55:00
```

Listing 4-13. Code to enable execution of disabled stored procedure

```
--ENABLE stored procedure for execution for NEXT EXECUTION

UPDATE FWDemo.ApplicationProcessProcedure
     SET Active = 1
WHERE ApplicationProcessProcedureID = 1    -- or value from Process Log
```

In another example, suppose the error occurred in the FWDemo.MonthlyProcess1 stored procedure. The process log, shown in Listing 4-14, shows that both of the daily process stored procedures completed successfully in the daily process controller.

Listing 4-14. Output from Process Log

```
ProcessLogID        ProcessLogMessage
CreateDate
79                  Procedure FWDemo.ProcessOrchestratorMD - ERROR
2018-09-18 19:35:00
78                  Procedure FWDemo.MonthlyProcessControllerMD - ERROR
2018-09-18 19:35:00
77                  Procedure FWDemo.MonthlyProcess1 - Problem Encountered
2018-09-18 19:35:00
76                  Procedure FWDemo.MonthlyProcess1 - STARTING
2018-09-18 19:34:00
75                  Executing FWDemo.MonthlyProcess1 ApplicatonID = 2
2018-09-18 19:34:00
                    ApplicatonProcessProcedureID = 3
74                  Procedure FWDemo.MonthlyProcessControllerMD - STARTING
2018-09-18 19:34:00
73                  Procedure FWDemo.DailyProcessControllerMD - COMPLETED
2018-09-18 19:34:00
72                  Procedure FWDemo.DailyProcess2 - COMPLETED
2018-09-18 19:34:00
71                  Procedure FWDemo.DailyProcess2 - STARTING
2018-09-18 19:33:00
70                  Executing FWDemo.DailyProcess2 ApplicatonID = 1
2018-09-18 19:33:00
```

```
                        ApplicatonProcessProcedureID = 2
69                      Procedure FWDemo.DailyProcess1 - COMPLETED
2018-09-18 19:33:00
68                      Procedure FWDemo.DailyProcess1 - STARTING
2018-09-18 19:32:00
67                      Executing FWDemo.DailyProcess1 ApplicatonID = 1
2018-09-18 19:32:00
                        ApplicatonProcessProcedureID = 1
66                      Procedure FWDemo.DailyProcessControllerMD - STARTING
2018-09-18 19:32:00
65                      Procedure FWDemo.ProcessOrchestratorMD - STARTING
2018-09-18 19:32:00
```

We see in the messages that their ApplicationProcessProcedureID values are 1 and 2, respectively. Code like that in Listing 4-15 could be used to manage the restart and subsequent process reset.

Listing 4-15. Manage error restart for FWDemo.MonthlyProcess1

```
--DISABLE stored procedures from execution for RESTART

UPDATE FWDemo.ApplicationProcessProcedure
     SET Active = 0
WHERE ApplicationProcessProcedureID in (1, 2)   -- or values from Process
Log
```

Listing 4-16 shows the log entries from the restarted application execution.

Listing 4-16. Process Log after restart

```
ProcessLogID      ProcessLogMessage
CreateDate
91                      Procedure FWDemo.ProcessOrchestratorMD - COMPLETED
2018-09-18 19:47:00
90                      Procedure FWDemo.MonthlyProcessControllerMD - COMPLETED
2018-09-18 19:47:00
89                      Procedure FWDemo.MonthlyProcess2 - COMPLETED
2018-09-18 19:47:00
```

```
88                 Procedure FWDemo.MonthlyProcess2 - STARTING
2018-09-18 19:46:00
87                 Executing FWDemo.MonthlyProcess2 ApplicatonID = 2
2018-09-18 19:46:00
                   ApplicatonProcessProcedureID = 4
86                 Procedure FWDemo.MonthlyProcess1 - COMPLETED
2018-09-18 19:46:00
85                 Procedure FWDemo.MonthlyProcess1 - STARTING
2018-09-18 19:45:00
84                 Executing FWDemo.MonthlyProcess1 ApplicatonID = 2
2018-09-18 19:45:00
                   ApplicatonProcessProcedureID = 3
83                 Procedure FWDemo.MonthlyProcessControllerMD - STARTING
2018-09-18 19:45:00
82                 Procedure FWDemo.DailyProcessControllerMD - COMPLETED
2018-09-18 19:45:00
81                 Procedure FWDemo.DailyProcessControllerMD - STARTING
2018-09-18 19:45:00
80                 Procedure FWDemo.ProcessOrchestratorMD - STARTING
2018-09-18 19:45:00
```

Once completed, the code shown in Listing 4-17 is used to reset the metadata for the application back to its original setting.

Listing 4-17. Enable stored procedures from execution for next execution

```
UPDATE FWDemo.ApplicationProcessProcedure
     SET Active = 1
WHERE ApplicationProcessProcedureID in (1, 2)   -- or values from Process Log
```

Initial setup of the metadata and deployment of new stored procedures to an application do, indeed, make the development and implementation of a process more complex. Hopefully, these metadata examples have demonstrated that management during execution, particularly in an error resolution situation, has become much simpler. Remember, setup and deployment may only occur once. Execution management will remain and occur over again as long as the process remains as part of the production activities. So, where do you want to make it easier?

This is the simplest use of metadata for our system and could be just the beginning. For a more advanced version, we could build a metadata layer on top of what we already have and let orchestrators execute using metadata for the controllers. But we will stop here because the idea has been established.

Summary

Metadata is used as the road map for the execution of the orchestrator and controllers. Most situations can now be handled by the manipulation of the metadata that controls the execution, and no manual deconstructions and executions are necessary. Any new stored procedures added to existing controllers or controllers added to existing orchestrators can be included simply by deploying the stored procedures to be compiled and adding the appropriate metadata. In most cases, if not all, no changes are required to existing stored procedures.

PART II

SSIS Frameworks

CHAPTER 5

A Simple, Custom, File-Based SSIS Framework

If you survey enterprises using SSIS for data integration/engineering, you will learn most enterprises do not use the SSIS catalog. Most enterprises execute SSIS from the file system. Why? Enterprises with a smaller number of SSIS packages outnumber enterprises with large numbers of SSIS packages. Executing a few dozen SSIS packages is different – really different – from managing the execution of thousands of SSIS packages. Don't take my word for it; ask any data engineer managing a larger enterprise.

Before the release of SSIS 2012 and the SSIS catalog, SSIS developers had to build their own SSIS frameworks. In this chapter, I share one way to build a custom file-based SSIS framework.

There are relatively few benefits to having 45 years' experience developing software. One benefit is living through several architecture pattern cycles, watching them wax and then wane in popularity. Old patterns get a fresh coat of virtual paint and become new again. (It reminds me of my older daughters asking me, in the 1990s, if I'd heard an awesome new band named Aerosmith.) We will see an example in a subsequent chapter when we deploy a remarkably similar SSIS framework to Azure Data Factory's Azure-SSIS integration runtime.

© Andy Leonard, Kent Bradshaw 2020
A. Leonard and K. Bradshaw, *SQL Server Data Automation Through Frameworks*,
https://doi.org/10.1007/978-1-4842-6213-9_5

An SSIS Framework, Defined and Designed

We author SSIS frameworks according to a few principles:

1. Functionality

2. Empathy

3. Simplicity

Functionality

In my opinion, an SSIS framework must accomplish execution, logging, and configuration. Execution of SSIS packages includes *grouped* execution. Execution grouping is the ability to execute a collection of SSIS packages in a specified order. The goal of logging is to surface enough operational information about an SSIS package execution to allow an experienced operator or developer to troubleshoot any execution error. Configuration promotes code reuse and supports SSIS Design Patterns by allowing developers and operators to configure SSIS package properties, parameters, and variables at runtime.

Empathy

"What do you mean by 'empathy,' Andy?" I am glad you asked. Empathy in software development shows up in user experience, or UX, design considerations. In SSIS design and development, empathy manifests by considering the skills of the user. In this case, the users consist of SSIS developers and operators. SSIS frameworks should employ the KISS ("keep it simple, stupid") principle. I assume the SSIS developer, or SSIS Developer Team, will be familiar with SSIS, so I build the SSIS framework *in* SSIS. I want the developer or team to manage and maintain their framework.

Simplicity

> *Software should be as complex as required to accomplish the goal, and not more complex than necessary.*

> —*Andy Leonard, circa 2007*

Managing complexity is hard for software developers. In the end, managing complexity is a balancing act between functionality, extensibility, and maintainability. An SSIS framework requires some amount of complexity to function, by its very nature. Supporting extensions via "extension points" always fails because framework authors cannot anticipate every use case in every enterprise. Designing any software – including an SSIS framework – should include thoughts about maintenance. Maintainability helps strike the balance between maximum functionality and minimum complexity.

Taken together, functionality, empathy, and simplicity are relatively straightforward concepts that are easy to comprehend but difficult to accomplish.

Building a File-Based SSIS Framework

In this portion of the book, over the next several chapters, we will build the components of an SSIS framework that will execute SSIS packages stored in the file system. The components of an SSIS framework are

- A database for execution and configuration metadata storage, log storage, and business logic

- An execution management engine built in SSIS

In this chapter, we focus on

- Building the metadata database

- Building a test SSIS project

- Adding metadata to the metadata database

Obtain the Code

To obtain the code for this book, visit the catalog page for this book on the Apress website (www.apress.com) or connect to the GitHub repository at

github.com/aleonard763/FrameworksBook

Save the code to a location you can readily access.

Metadata-Driven Execution Management

Metadata-driven execution management is not a complex topic even though it may appear complex at first blush. The example to follow builds a database named SSISConfig and the execution metadata tables shown in Figure 5-1.

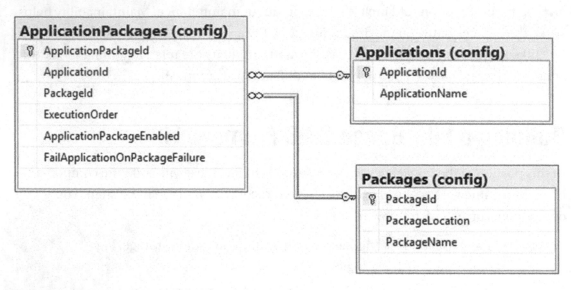

Figure 5-1. *Metadata tables for execution management*

The SSISConfig Database

The SSISConfig database is designed to contain metadata for the framework.

   Create the SSISConfig database using the T-SQL shown in Listing 5-1.

Listing 5-1. Creating the SSISConfig database

```
Use [master]
go

print 'SSISConfig database'
If Not Exists(Select [databases].[name]
             From [sys].[databases]
             Where [databases].[name] = N'SSISConfig')
  begin
```

```
  print ' - Create SSISConfig database'
  Create Database SSISConfig
  print ' - SSISConfig database created'
 end
Else
 begin
  print ' - SSISConfig database already exists.'
 end
print ''
go
```

SQL Server Management Studio (SSMS) Object Explorer should appear similar to Figure 5-2.

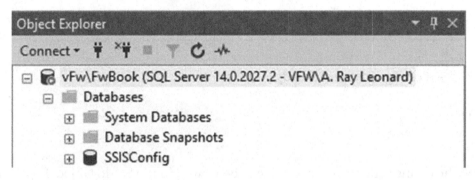

Figure 5-2. *SSISConfig database created*

Config Schema

At the top of the list of software development best practices is "separation of concerns." One way to separate concerns is decoupling. One can argue database development is not software development. Others will argue some software development best practices may – perhaps *should* – be used in database development. I am in this second camp.

Add the first schema – config – to the SSISConfig database using the T-SQL shown in Listing 5-2.

Listing 5-2. Creating the config schema

```
use [SSISConfig]
go

print 'Config schema'
If Not Exists(Select [schemas].[name]
              From [sys].[schemas]
              Where [schemas].[name] = N'config')
 begin
  print ' - Create config schema'
  declare @sql nvarchar(100) = N'Create Schema config'
  exec(@sql)
  print ' - Config schema created'
 end
Else
 begin
  print ' - Config schema already exists.'
 end
print ''
go
```

Once the config schema is created, it can be viewed in the SSMS Object Explorer. Navigate to the SSISConfig ➤ Security ➤ Schemas node as shown in Figure 5-3.

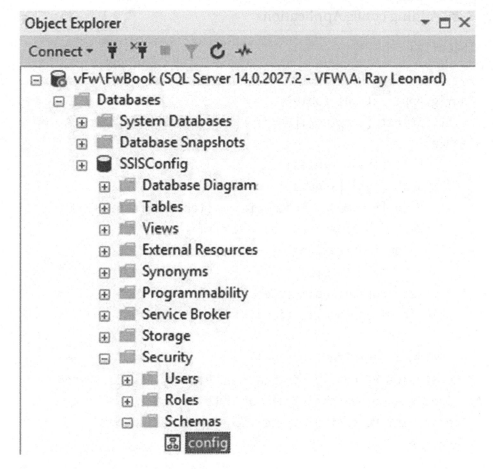

Figure 5-3. *Viewing the config schema*

The Config.Applications Table

In this framework, an *SSIS application* is defined as "a collection of SSIS packages configured to execute in a prescribed order." Each application is identified by an ApplicationId (non-null int).

Each application has a name attribute, and the ApplicationName must be unique. Identifying the application by ApplicationId is important for referential integrity. Observing the T-SQL for tables in the config schema – especially the config. ApplicationPackages table – you will note the design is third normal form.

Execute the T-SQL in Listing 5-3 to add the Applications table to the config schema in the SSISConfig database.

Listing 5-3. Adding config.Applications

```
use [SSISConfig]
go

print 'Config.Applications table'
If Not Exists(Select [schemas].[name] + '.' + [tables].[name] As ➤
[Schema.Table]
              From [sys].[tables]
              Join [sys].[schemas]
                On [schemas].[schema_id] = [tables].[schema_id]
              Where [schemas].[name] = N'config'
                And [tables].[name] = N'Applications')
 begin
  print ' - Create config.Applications table'
  Create Table [config].[Applications]
   (
      ApplicationId int identity(1, 1)
        Constraint PK_config_Applications Primary Key Clustered
    , ApplicationName nvarchar(255) Not NULL
        Constraint UQ_config_Applications_ApplicationName
          Unique
   )
  print ' - Config.Applications table created'
 end
Else
 begin
  print ' - Config.Applications table already exists.'
 end
print ''
go
```

Once executed, the config.Applications table appears in Object Explorer as shown in Figure 5-4.

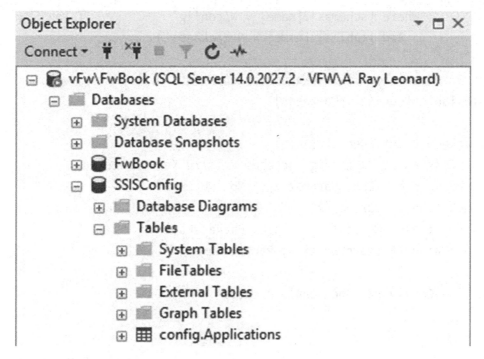

Figure 5-4. *Config.Applications table*

The Config.Packages Table

Package location metadata is stored in the config.Packages table. As we build out other versions of this SSIS framework, we will see this is the table that changes. The changes are based on where SSIS packages reside in the enterprise.

Create the config.Packages table by executing the T-SQL shown in Listing 5-4.

Listing 5-4. Creating the config.Packages table

```
use [SSISConfig]
go

print 'Config.Packages table'
If Not Exists(Select [schemas].[name] + '.' + [tables].[name] As ➤
[Schema.Table]
            From [sys].[tables]
            Join [sys].[schemas]
              On [schemas].[schema_id] = [tables].[schema_id]
```

```
            Where [schemas].[name] = N'config'
              And [tables].[name] = N'Packages')
 begin
  print ' - Create config.Packages table'
  Create Table [config].[Packages]
   (
      PackageId int identity(1, 1)
         Constraint PK_config_Packages Primary Key Clustered
      , PackageLocation nvarchar(255) Not NULL
    , PackageName nvarchar(255) Not NULL
      , Constraint UQ_config_Packages_PackageName
          Unique(PackageLocation, PackageName)
   )
  print ' - Config.Packages table created'
 end
Else
 begin
  print ' - Config.Packages table already exists.'
 end
print ''
go
```

Once created, the config.Packages table appears in Object Explorer as shown in Figure 5-5.

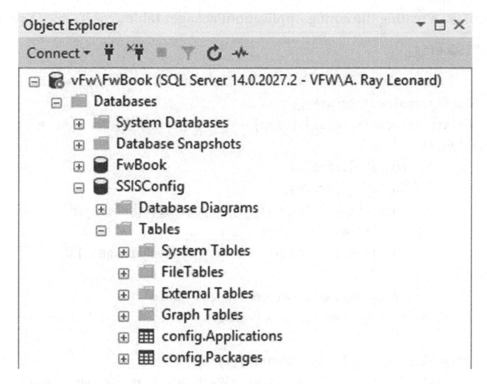

Figure 5-5. *Creating config.Packages table*

As mentioned earlier, an SSIS application in the framework is a collection of packages configured to execute in a specific order. Considering the cardinality of applications to packages, it's obvious the answer is one to many, at least in part.

Consider utility SSIS packages, such as a package that archives flat files after the data contained in them has been successfully loaded or staged in a database. Such a package – perhaps named ArchiveFile.dtsx – could be parameterized with parameters like SourceFilePath and DestinationLocation. ArchiveFile.dtsx could then be reused in several SSIS applications.

The cardinality of these applications to the ArchiveFile.dtsx package is many-to-one. Combining the cardinalities one-to-many and many-to-one, one gets many-to-many. Resolving many-to-many in a third normal form design requires an additional table: config.ApplicationPackages in this case. Create config.ApplicationPackages using the T-SQL shown in Listing 5-5.

Listing 5-5. Creating the config.ApplicationPackages table

```
use [SSISConfig]
go

print 'Config.ApplicationPackages table'
If Not Exists(Select [schemas].[name] + '.' + [tables].[name] As ➤
[Schema.Table]
             From [sys].[tables]
             Join [sys].[schemas]
               On [schemas].[schema_id] = [tables].[schema_id]
             Where [schemas].[name] = N'config'
               And [tables].[name] = N'ApplicationPackages')
 begin
  print ' - Create config.ApplicationPackages table'
  Create Table [config].[ApplicationPackages]
   (
     ApplicationPackageId int identity(1, 1)
        Constraint PK_config_ApplicationPackages Primary Key Clustered
     , ApplicationId int Not NULL
        Constraint FK_config_ApplicationPackages_config_Applications
          Foreign Key References [config].[Applications](ApplicationId)
     , PackageId int Not NULL
        Constraint FK_config_ApplicationPackages_config_Packages
          Foreign Key References [config].[Packages](PackageId)
     , ExecutionOrder int Not NULL
        Constraint DF_config_ApplicationPackages_ExecutionOrder
            Default(10)
     , ApplicationPackageEnabled bit Not NULL
        Constraint DF_config_ApplicationPackages_ApplicationPackageEnabled
            Default(1)
     , FailApplicationOnPackageFailure bit Not NULL
       Constraint ➤ DF_config_ApplicationPackages_
       FailApplicationOnPackageFailure
          Default(1)
```

```
  , Constraint ➤ UQ_config_ApplicationPackages_ApplicationId_
  PackageId_ExecutionOrder
      Unique(ApplicationId, PackageId, ExecutionOrder)
  )
 print ' - Config.ApplicationPackages table created'
 end
Else
 begin
  print ' - Config.ApplicationPackages table already exists.'
 end
print ''
go
```

When the config.ApplicationPackages is created, Object Explorer shows the SSISConfig tables as shown in Figure 5-6.

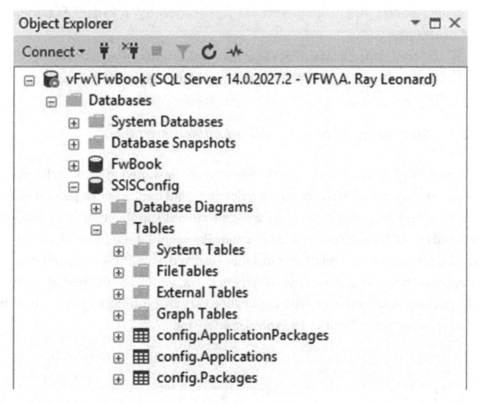

Figure 5-6. *The config.ApplicationPackages table*

Examine the config.ApplicationPackages table in Object Explorer by expanding the Columns, Keys, and Constraints virtual folders as shown in Figure 5-7.

Figure 5-7. *Examining the config.ApplicationPackages table*

The config.ApplicationPackages table columns are designed to resolve the many-to-many cardinality between the config.Applications and config.Packages tables. Each entry maps SSIS package metadata (via PackageId) to SSIS application metadata (via ApplicationId) to an identity column, ApplicationPackageId, thereby *resolving* (or *bridging*) the many-to-many relationship. Once resolved, this SSISConfig database design supports the execution of *Application Packages*, which are defined as an instance of an SSIS package mapped to an SSIS application. Application Packages are then, by design, the smallest unit of work in this SSIS framework.

Attributes of Execution

One implication of Application Packages being the smallest unit of work is Application Packages are where we define *attributes of execution*. The application package attributes of execution in this framework are

- ExecutionOrder

- ApplicationPackageEnabled

- FailApplicationOnPackageFailure

Execution Order is an integer that represents the "in a specified order" part of the SSIS application definition.

Application Package Enabled is a bit field that indicates whether the individual application package is configured to execute when the SSIS application is executed.

Fail Application On Package Failure is a bit that configures fault tolerance. Consider what happens when one executes an SSIS application: a number of SSIS packages are executed in a specified order. SSIS packages with a specified order within an SSIS application are application packages. What if we would like it very much if an SSIS package executed, completed, and succeeded, but if the package fails for some reason, it's not vital for this particular SSIS application execution? What if we just don't care so much if this particular package fails? The FailApplicationOnPackageFailure attribute will permit the package to fail *without* stopping the SSIS application execution.

Default Constraints

A default constraint manages the initial value of a column when a new record is added to a table and the value for field is not supplied an explicit value:

- DF_config_ApplicationPackages_FailApplicationOnPackageFailure
  is a default constraint that sets the value of the
  FailApplicationOnPackageFailure bit to True (1) if no value is
  supplied when the config.ApplicationPackages record is first added
  to the table.

- DF_config_ApplicationPackages_ApplicationPackageEnabled
  is a default constraint that sets the value of the
  ApplicationPackageEnabled bit to True (1) if no value is supplied
  when the config.ApplicationPackages record is first added to the table.

73

- DF_config_ApplicationPackages_ExecutionOrder is a default constraint that sets the value of the ExecutionOrder integer to 10 if no value is supplied when the config.ApplicationPackages record is first added to the table.

Relationships

Keys manage relationships between entities in the SSISConfig database and therefore in the SSIS framework:

- Config.ApplicationPackages is configured with PK_config_ ApplicationPackages. Each table has a primary key for relationship management. In this framework, PK_config_ApplicationPackages serves to help identify each Application-Package-Execution Order combination.

- FK_config_ApplicationPackages_config_Applications is a foreign key between the config.ApplicationPackages.ApplicationId field and the config.Applications.ApplicationId field. Before a row is added to config.ApplicationPackages, a corresponding row *must* already exist in the config.Applications table. The Application is mapped to a Package that executes at this *position* – denoted by the value of ExecutionOrder – in the application.

- FK_config_ApplicationPackages_config_Packages is a foreign key between the config.ApplicationPackages.PackageId field and the config.Packages.PackageId field. Before a row is added to config. ApplicationPackages, a corresponding row *must* already exist in the config.Packages table. A Package that executes at a *position* – denoted by the value of ExecutionOrder – is mapped into an Application.

- UQ_config_ApplicationPackages_ApplicationId_PackageId_ ExecutionOrder is a unique constraint that guarantees the distinctness – across the ApplicationId, PackageId, and ExecutionOrder fields – of each application package stored in the config.ApplicationPackages table. This means two (or more) records in the config.ApplicationPackages table *may not have identical* ApplicationId, PackageId, and ExecutionOrder field values.

Frequently Asked Questions

Frameworks are new to many data users, and there are several questions we hear regularly:

- Is it possible for two (or more) records in the config. ApplicationPackages table to have the same ApplicationId? Yes. In fact, the framework is designed for this use case. Several config. ApplicationPackages records with the same ApplicationId value is how the framework is used to configure an SSIS application.

- Is it possible for two (or more) records in the config. ApplicationPackages table to have the same PackageId? Yes. This supports code reuse, such as executing the ArchiveFile.dtsx SSIS package in several applications. This also supports an SSIS Design Pattern called "Range-Based Load." In a Range-Based Load, the *same* SSIS package executes several times. New minimum and maximum range values, or other mathematical functions, are passed to the package for each execution. For example, the same SSIS package could execute ten times. The first execution could load records in which a numeric value begins with the numeral "0." Subsequent executions could load data where the numeric value starts with the numerals "1," "2," and so on.

We will return to SSMS and development of the SSISConfig database in a bit. Let's next build a sample SSIS solution to use to test the framework.

A Sample SSIS Solution

Testing is hard. Testing requires thinking, differently, about how users will interact with an application. Developers are notoriously bad at testing their own code. Why? I believe it's because development focuses on delivering functionality that just works. For some (like me), it's difficult to change gears and begin thinking about all the permutations a user may enter – or *forget* to enter – when interacting with our code.

A first thought for building a sample SSIS solution for testing is, "I can do this with one SSIS package." That's incorrect. Testing requires at least two SSIS packages, one package that succeeds and another that fails.

Configure Visual Studio 2019 for SSIS Development

Before you build an SSIS 2019 sample solution, you need to download and install Visual Studio 2019. To begin, browse to visualstudio.microsoft.com and select a version of Visual Studio 2019 to install, as shown in Figure 5-8.

Figure 5-8. *Preparing to download Visual Studio 2019*

For the purposes of the example used in this book, the Community 2019 edition will suffice (plus, it's free).

The next step is to install the Integration Services extension. Visual Studio is an *integrated development environment,* or IDE. The IDE serves as a host, or shell, for various and sundry templates that enable software development experiences in different languages and platforms.

Before Visual Studio 2019, SSIS templates were installed by executing a SQL Server Data Tools, or SSDT, stand-alone installer. Starting with Visual Studio 2019, SSIS templates are managed in the Visual Studio Marketplace as another available Visual Studio extension.

To find and install the Integration Services extension in Visual Studio 2019, open Visual Studio and click Extensions ➤ Manage Extensions, as shown in Figure 5-9.

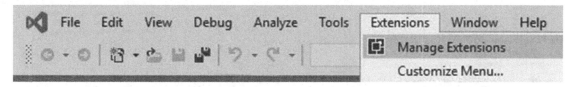

Figure 5-9. *Preparing to install the Integration Services extension*

When the Manage Extensions dialog displays, search for "Integration Services" in the search textbox. Select the "SQL Server Integration Services Projects" extension, and then click the Download button, as shown in in Figure 5-10.

Figure 5-10. *Downloading the "SQL Server Integration Services Projects" extension*

When the new extension is downloaded, follow the instructions presented from the Visual Studio 2019 extension process. The extension installation process likely begins with closing the Visual Studio IDE. When the extension installation is complete, Visual Studio 2019 may be used to develop SSIS projects.

Create the Sample SSIS Solution

To begin, create a new SSIS solution named TestSSISSolution – containing a new SSIS project named TestSSISProject – as shown in Figure 5-11.

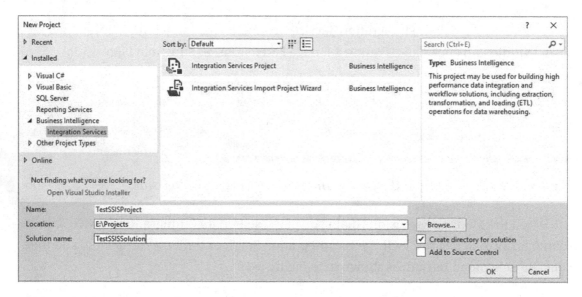

Figure 5-11. *Creating the TestSSISProject in TestSSISSolution*

When the project opens, rename the default SSIS package to "ReportAndSucceed. dtsx" as shown in Figure 5-12.

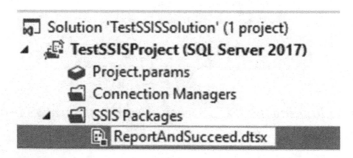

Figure 5-12. *Renaming the package "ReportAndSucceed.dtsx"*

Add a Script Task to the ReportAndSucceed's Control Flow by dragging a Script Task from the SSIS Toolbox onto the Control canvas, and then rename it to "SCR Log Values," as shown in Figure 5-13.

Figure 5-13. *Adding and renaming "SCR Log Values" script task*

Open the Script Task Editor by right-clicking the "SCR Log Values" script task and clicking "Edit." When the "SCR Log Values" script task editor displays, add the following variables to the ReadOnlyVariables property list:

- System::PackageName

- System::TaskName

The Script Task Editor will appear as shown in Figure 5-14.

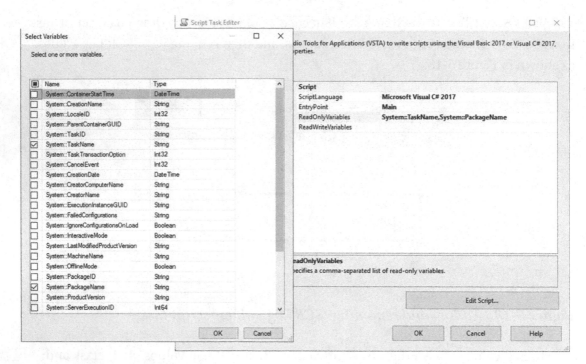

Figure 5-14. *Adding System::PackageName and System::TaskName to ReadOnlyVariables*

Click the Edit Script button, and enter the C# shown in Listing 5-6 into the public void Main() method.

Listing 5-6. Main() method C# to configure raising an Information event in the ReportAndSucceed SSIS package

```
public void Main()
        {
        string packageName = ➤ Dts.Variables["System::PackageName"].
        Value.ToString();
        string taskName = ➤ Dts.Variables["System::TaskName"].Value.
        ToString();
        string subComponent = packageName + "." + taskName;
        int informationCode = 1001;
        bool fireAgain = true;
        string description = "I am " + packageName;
```

```
    Dts.Events.FireInformation(informationCode, subComponent, ➤
    description, "", 0, ref fireAgain);

    Dts.TaskResult = (int)ScriptResults.Success;
    }
```

When the code has been entered into the Main() method, it will appear as shown in Figure 5-15.

```
public void Main()
{
    string packageName = Dts.Variables["System::PackageName"].Value.ToString();
    string taskName = Dts.Variables["System::TaskName"].Value.ToString();
    string subComponent = packageName + "." + taskName;
    int informationCode = 1001;
    bool fireAgain = true;
    string description = "I am " + packageName;

    Dts.Events.FireInformation(informationCode, subComponent, description, "", 0, ref fireAgain);

    Dts.TaskResult = (int)ScriptResults.Success;
}
```

Figure 5-15. *Main() method C# that raises an Information event*

Close the Script editor window, and click the OK button on the SCR Log Values script task. Click Debug ➤ Start Debugging as shown in Figure 5-16.

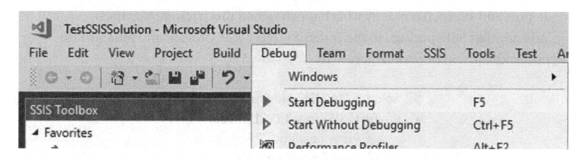

Figure 5-16. *Starting a debug execution of the ReportAndSucceed package*

When execution completes – and hopefully succeeds – click the Progress (or Execution Results, if you stop the debugger) to view the Information event we configure in the SCR Log Values script task as shown in Figure 5-17.

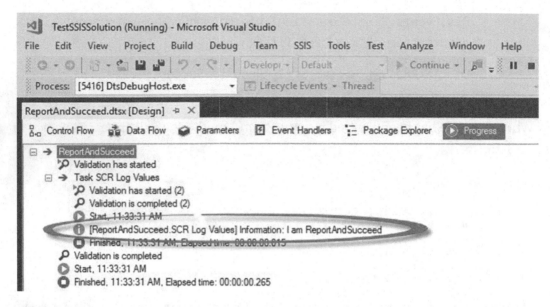

Figure 5-17. *The Information event*

The Information event reads: "[ReportAndSucceed.SCR Log Values] Information: I am ReportAndSucceed".

The first part of the message is the subComponent and is enclosed in square brackets ("[]"). The subComponent is built from the Package name and Task name variables: "ReportAndSucceed.SCR Log Values". The event name – Information – follows the subComponent. The description is next and reads "I am ReportAndSucceed".

Report and succeed is exactly what happens when this package executes.

Add another SSIS package to the TestSSISProject SSIS project, and rename it to "ReportAndFail," as shown in Figure 5-18.

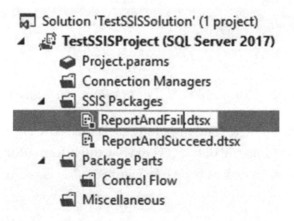

Figure 5-18. *Adding the ReportAndFail SSIS package*

As before, add a Script Task named "SCR Log Values" to the ReportAndFail package. Add the System::PackageName and System::TaskName variables to the ReadOnlyVariables property, and then click the Edit Script button. Add the C# in Listing 5-7 to the Main() method.

Listing 5-7. Main() method C# to configure and raise an Error event in the ReportAndFail SSIS package

```
public void Main()
          {
          string packageName = ➤ Dts.Variables["System::PackageName"].
          Value.ToString();
          string taskName = ➤ Dts.Variables["System::TaskName"].Value.
          ToString();
          string subComponent = packageName + "." + taskName;
          int errorCode = -1001;
          string description = packageName + " execution failed";

          Dts.Events.FireError(errorCode, subComponent, description,
          "", ➤ 0);

          Dts.TaskResult = (int)ScriptResults.Success;
          }
```

When the code has been entered into the Main() method, it will appear as shown in Figure 5-19.

```
public void Main()
{
    string packageName = Dts.Variables["System::PackageName"].Value.ToString();
    string taskName = Dts.Variables["System::TaskName"].Value.ToString();
    string subComponent = packageName + "." + taskName;
    int errorCode = -1001;
    string description = packageName + " execution failed";

    Dts.Events.FireError(errorCode, subComponent, description, "", 0);

    Dts.TaskResult = (int)ScriptResults.Success;
}
```

Figure 5-19. *Main() method C# that raises an Error event*

Close the Script editor window by clicking the "X" in the upper-right corner of the VstaProjects .Net code editor, or by clicking File ➤ Exit, and then click the OK button on the SCR Log Values script task. Click Debug ➤ Start Debugging to start debug execution of the ReportAndFail SSIS package.

When execution completes – and hopefully *fails* – click the Progress (or Execution Results, if you stop the debugger) to view the Information event we configure in the SCR Log Values script task as shown in Figure 5-20.

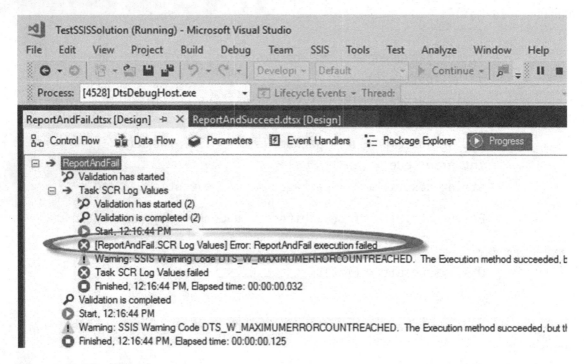

Figure 5-20. *The Error event*

The error event reads: "[ReportAndFail.SCR Log Values] Error: ReportAndFail execution failed".

The first part of the message is the subComponent and is enclosed in square brackets ("[]"). The subComponent is built from the Package name and Task name variables: "ReportAndFail.SCR Log Values". The event name – Error – follows the subComponent. The description is next and reads "ReportAndFail execution failed".

Report and fail is exactly what happens when this package executes.

At a minimum, a test project should include components that succeed *and* fail.

SSIS Framework Metadata Management

In enterprise architecture or any flavor of development, solving one problem often creates a new problem, or problems. I call this "there's no free lunch." Growing through the "there's no free lunch" phase is a rite of passage for developers, it's a natural part of maturing, and it's a *requirement* for technology architects.

For example, the SSIS framework described in this chapter will simplify the execution of several SSIS packages in a specified order by using metadata. There will be lots of metadata to manage.

There's no free lunch.

All people who write T-SQL have preferences regarding capitalization and indentation. I am no different. I find my T-SQL capitalization and indentation helps me think about the problem I am trying to solve.

Asking yourself or your team, "What is the problem I am/we are trying to solve?" is potent. Try it and see.

Add an SSIS Application

Add an SSIS application to the SSIS framework by executing the T-SQL shown in Listing 5-8.

Listing 5-8. Adding an SSIS application

```
use [SSISConfig]
go

Set NoCount ON

declare @ApplicationName nvarchar(255) = N'Framework Test'

print @ApplicationName
declare @ApplicationId int = (Select [Applications].[ApplicationId]
                              From [config].[Applications]
                              Where [Applications].[ApplicationName] =
                              ➤ @ApplicationName)
```

```
If (@ApplicationId Is NULL)
 begin
  print ' - Adding ' + @ApplicationName + ' application to ➤
  config.Applications table'

  declare @AppTbl table(ApplicationId int)

  Insert Into [config].[Applications]
  (ApplicationName)
  Output inserted.ApplicationId into @AppTbl
  Values (@ApplicationName)

  Set @ApplicationId = (Select ApplicationId
                       From @AppTbl)

  print ' - ' + @ApplicationName + ' application added to ➤
  config.Applications table'
 end
Else
 begin
  print ' - ' + @ApplicationName + ' application already exists in ➤
  the config.Applications table.'
 end

 Select @ApplicationId As ApplicationId
print ''
```

The T-SQL in Listing 5-8 is *idempotent*. Idempotence is a mathematical term that means an operation may be applied multiple times and produce the same result. Idempotence is another way of saying "re-executable code."

Applied to the T-SQL code in Listing 5-8, re-executable code produces the same result in the config.Applications table; metadata for a new SSIS application named "Framework Test" is added to the framework. The first execution of the T-SQL in Listing 5-8 returns the messages (to the Messages tab in SSMS) shown in Listing 5-9.

Listing 5-9. Messages returned when adding the Framework Test SSIS application for the first time

```
Framework Test
 - Adding Framework Test application to config.Applications table
 - Framework Test application added to config.Applications table
```

The result, which is the ApplicationId for the "Framework Test" SSIS application, is shown in Figure 5-21.

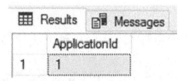

Figure 5-21. *"Framework Test" ApplicationId*

Re-executing the Add SSIS Application T-SQL shown in Listing 5-8 produces the messages shown in Figure 5-22.

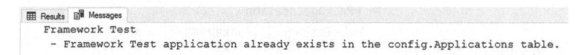

Figure 5-22. *Re-executing SSIS Application T-SQL*

Re-executing the Add SSIS Application T-SQL shown in Listing 5-8 produces the *same* results, as shown in Figure 5-23.

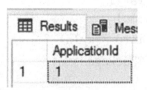

Figure 5-23. *Identical Framework Test ApplicationId*

Subsequent re-executions of the Add SSIS Application T-SQL produce the same result.

Add SSIS Packages

Add SSIS packages in TestSSISProject to the SSIS framework by executing the T-SQL shown in Listing 5-10.

Listing 5-10. Adding SSIS packages from TestSSISProject

```
use [SSISConfig]
go

Set NoCount ON

declare @PackageLocation nvarchar(255) = ➤ N'E:\Projects\TestSSISSolution\
TestSSISProject\'
declare @PackageName nvarchar(255) = N'ReportAndSucceed.dtsx'

print @PackageLocation + @PackageName
declare @PackageId int = (Select [Packages].[PackageId]
                          From [config].[Packages]
                          Where [Packages].[PackageLocation] =
                          ➤ @PackageLocation
                            And [Packages].[PackageName] = @PackageName)
If (@PackageId Is NULL)
 begin
  print ' - Adding ' + @PackageName + ' package to config.Packages table'

  declare @PkgTbl table(PackageId int)

  Insert Into [config].[Packages]
  (PackageLocation, PackageName)
  Output inserted.PackageId into @PkgTbl
  Values (@PackageLocation, @PackageName)

  Set @PackageId = (Select PackageId
                      From @PkgTbl)

  print ' - ' + @PackageName + ' package added to config.Packages table'
 end
Else
```

```
 begin
  print ' - ' + @PackageName + ' package already exists in the ➤ config.
  Packages table.'
 end

 Select @PackageId As PackageId
print ''

set @PackageName = N'ReportAndFail.dtsx'

print @PackageLocation + @PackageName
set @PackageId = (Select [Packages].[PackageId]
                  From [config].[Packages]
                  Where [Packages].[PackageLocation] = @PackageLocation
                    And [Packages].[PackageName] = @PackageName)
If (@PackageId Is NULL)
 begin
  print ' - Adding ' + @PackageName + ' application to config.Packages
  table'

  Delete @PkgTbl

  Insert Into [config].[Packages]
  (PackageLocation, PackageName)
  Output inserted.PackageId into @PkgTbl
  Values (@PackageLocation, @PackageName)

  Set @PackageId = (Select PackageId
                    From @PkgTbl)

  print ' - ' + @PackageName + ' package added to config.Packages table'
 end
Else
 begin
  print ' - ' + @PackageName + ' package already exists in the ➤ config.
  Packages table.'
 end

 Select @PackageId As PackageId
print ''
```

Like the T-SQL in Listing 5-8, the T-SQL in Listing 5-10 is also idempotent, so the T-SQL code in Listing 5-10 produces the same result in the config.Packages table; metadata for two new SSIS packages named "ReportAndSucceed.dtsx" and "ReportAndFail.dtsx" are added to the framework. The first execution of the T-SQL in Listing 5-10 returns the messages (to the Messages tab in SSMS) shown in Listing 5-11.

Listing 5-11. Messages returned when adding ReportAndSucceed.dtsx and ReportAndFail.dtsx SSIS packages for the first time

```
E:\Projects\TestSSISSolution\TestSSISProject\ReportAndSucceed.dtsx
  - Adding ReportAndSucceed.dtsx package to config.Packages table
  - ReportAndSucceed.dtsx package added to config.Packages table

E:\Projects\TestSSISSolution\TestSSISProject\ReportAndFail.dtsx
  - Adding ReportAndFail.dtsx application to config.Packages table
  - ReportAndFail.dtsx package added to config.Packages table
```

The result, which is the PackageIds for the "ReportAndSucceed.dtsx" and "ReportAndFail.dtsx" SSIS packages, is shown in Figure 5-24.

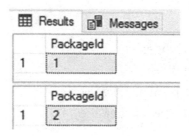

Figure 5-24. *"ReportAndSucceed.dtsx" and "ReportAndFail.dtsx" SSIS PackageIds*

Re-executing the Add SSIS Packages T-SQL shown in Listing 5-10 produces the messages shown in Figure 5-25.

```
 Results  Messages
    E:\Projects\TestSSISSolution\TestSSISProject\ReportAndSucceed.dtsx
     - ReportAndSucceed.dtsx package already exists in the config.Packages table.

    E:\Projects\TestSSISSolution\TestSSISProject\ReportAndFail.dtsx
     - ReportAndFail.dtsx package already exists in the config.Packages table.
```

Figure 5-25. *Re-executing SSIS Packages T-SQL*

Re-executing the Add SSIS Packages T-SQL shown in Listing 5-10 produces the *same* results, as shown in Figure 5-26.

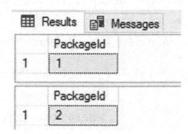

Figure 5-26. *Identical PackageIds for "ReportAndSucceed.dtsx" and "ReportAndFail.dtsx" SSIS packages*

Subsequent re-executions of the Add SSIS Packages T-SQL produce the same result.

Assign SSIS Application Packages

Assign SSIS application packages in SSIS framework by executing the T-SQL shown in Listing 5-12.

Listing 5-12. Assigning SSIS application packages

```
use [SSISConfig]
go

Set NoCount ON

declare @ApplicationName nvarchar(255) = N'Framework Test'
declare @PackageLocation nvarchar(255) = ➤ N'E:\Projects\TestSSISSolution\
TestSSISProject\'
declare @PackageName nvarchar(255) = N'ReportAndSucceed.dtsx'
declare @ExecutionOrder int = 10

print @ApplicationName + ' - ' + @PackageLocation + @PackageName

declare @ApplicationId int = (Select [Applications].[ApplicationId]
                              From [config].[Applications]
                              Where [Applications].[ApplicationName] =
                              ➤ @ApplicationName)
```

```
declare @PackageId int = (Select [Packages].[PackageId]
                          From [config].[Packages]
                          Where [Packages].[PackageLocation] =
                        ➤ @PackageLocation
                            And [Packages].[PackageName] = @PackageName)

declare @ApplicationPackageId int = (Select ApplicationPackageId
                                     From config.ApplicationPackages
                                     Where ApplicationId = @ApplicationId
                                       And PackageId = @PackageId
                                       And ExecutionOrder =
                                       @ExecutionOrder)

If (@ApplicationPackageId Is NULL)
 begin
   print ' - Assigning ' + @PackageName + ' package to '
        + @ApplicationName + ' application'
        + ' in config.ApplicationPackages table'
        + ' at ExecutionOrder ' + Convert(varchar(9), @ExecutionOrder)

   Insert Into [config].[ApplicationPackages]
   (ApplicationId
 , PackageId
 , ExecutionOrder)
   Values (@ApplicationId
         , @PackageId
         , @ExecutionOrder)

   print ' - ' + @PackageName + ' package assigned to '
        + @ApplicationName + ' application'
        + ' in config.ApplicationPackages table'
        + ' at ExecutionOrder ' + Convert(varchar(9), @ExecutionOrder)
 end
Else
 begin
print ' - ' + @PackageName + ' package already'
     + ' assigned to ' + @ApplicationName
```

```
      + ' application in config.ApplicationPackages table'
      + ' at ExecutionOrder ' + Convert(varchar(9), @ExecutionOrder)
      + '.'
 end
print ''

set @PackageName = N'ReportAndFail.dtsx'
set @ExecutionOrder = 20

print @ApplicationName + ' - ' + @PackageLocation + @PackageName

set @ApplicationId = (Select [Applications].[ApplicationId]
                         From [config].[Applications]
                         Where [Applications].[ApplicationName] =
                         ➤ @ApplicationName)

set @PackageId = (Select [Packages].[PackageId]
                  From [config].[Packages]
                  Where [Packages].[PackageLocation] = @PackageLocation
                    And [Packages].[PackageName] = @PackageName)

set @ApplicationPackageId = (Select ApplicationPackageId
                             From config.ApplicationPackages
                             Where ApplicationId = @ApplicationId
                               And PackageId = @PackageId
                               And ExecutionOrder = @ExecutionOrder)

If (@ApplicationPackageId Is NULL)
 begin
   print ' - Assigning ' + @PackageName + ' package to '
       + @ApplicationName + ' application'
       + ' in config.ApplicationPackages table'
       + ' at ExecutionOrder ' + Convert(varchar(9), @ExecutionOrder)

   Insert Into [config].[ApplicationPackages]
   (ApplicationId
 , PackageId
 , ExecutionOrder)
   Values (@ApplicationId
```

```
      , @PackageId
      , @ExecutionOrder)

 print ' - ' + @PackageName + ' package assigned to '
      + @ApplicationName + ' application'
      + ' in config.ApplicationPackages table'
      + ' at ExecutionOrder ' + Convert(varchar(9), @ExecutionOrder)
 end
Else
 begin
  print ' - ' + @PackageName + ' package already'
      + ' assigned to ' + @ApplicationName
      + ' application in config.ApplicationPackages table'
      + ' at ExecutionOrder ' + Convert(varchar(9), @ExecutionOrder)
      + '.'
 end
print ''
```

Like the T-SQL in Listing 5-8 and Listing 5-10, the T-SQL in Listing 5-12 is idempotent. The T-SQL code in Listing 5-12 produces the same result in the config.ApplicationPackages table; metadata for two new SSIS packages named "ReportAndSucceed.dtsx" and "ReportAndFail.dtsx" are assigned to the "Framework Test" application in the framework. The first execution of the T-SQL in Listing 5-12 returns the messages shown in Listing 5-13.

Listing 5-13. Messages returned when assigning ReportAndSucceed.dtsx and ReportAndFail.dtsx SSIS packages to the Framework Test application for the first time

```
Framework Test - ➤ E:\Projects\TestSSISSolution\TestSSISProject\
ReportAndSucceed.dtsx
 - Assigning ReportAndSucceed.dtsx package to Framework Test application in
 ➤ config.ApplicationPackages table at ExecutionOrder 10
 - ReportAndSucceed.dtsx package assigned to Framework Test application in
 ➤ config.ApplicationPackages table at ExecutionOrder 10
```

```
Framework Test - ➤ E:\Projects\TestSSISSolution\TestSSISProject\
ReportAndFail.dtsx
 - Assigning ReportAndFail.dtsx package to Framework Test application in ➤
config.ApplicationPackages table at ExecutionOrder 20
 - ReportAndFail.dtsx package assigned to Framework Test application in ➤
config.ApplicationPackages table at ExecutionOrder 20
```

Re-executing the Add SSIS Application Packages T-SQL shown in Listing 5-12 produces the messages shown in Figure 5-27.

```
Messages
   Framework Test - E:\Projects\TestSSISSolution\TestSSISProject\ReportAndSucceed.dtsx
    - ReportAndSucceed.dtsx package already assigned to Framework Test application in config.ApplicationPackages table at ExecutionOrder 10.

   Framework Test - E:\Projects\TestSSISSolution\TestSSISProject\ReportAndFail.dtsx
    - ReportAndFail.dtsx package already assigned to Framework Test application in config.ApplicationPackages table at ExecutionOrder 20.
```

Figure 5-27. *Re-executing SSIS Application Packages T-SQL*

Subsequent re-executions of the Add SSIS Application Packages T-SQL produce the same result.

The SSIS framework metadata database contains a schema named config. The config schema contains three tables:

1. Config.Applications

2. Config.Packages

3. Config.ApplicationPackages

The tables contain metadata for an application named Framework Test, which is related to two SSIS packages named ReportAndSucceed.dtsx and ReportAndFail.dtsx. It's possible to retrieve the metadata by executing the T-SQL shown in Listing 5-14.

Listing 5-14. Viewing the framework application contents

```
Use [SSISConfig]
go

declare @ApplicationName nvarchar(255) = N'Framework Test'

Select a.ApplicationName
    , p.PackageLocation + p.PackageName As PackagePath
    , ap.ExecutionOrder
    , ap.FailApplicationOnPackageFailure
```

```
From [config].[ApplicationPackages] ap
Join [config].[Applications] a
  On a.ApplicationId = ap.ApplicationId
Join [config].Packages p
  On p.PackageId = ap.PackageId
Where a.ApplicationName = @ApplicationName
  And ap.ApplicationPackageEnabled = 1
Order By ap.ExecutionOrder
```

The SSIS framework metadata returned from the T-SQL query shown in Listing 5-14 is shown in Figure 5-28.

	ApplicationName	PackagePath	ExecutionOrder	FailApplicationOnPackageFailure
1	Framework Test	E:\Projects\TestSSISSolution\TestSSISProject\ReportAndSucceed.dtsx	10	1
2	Framework Test	E:\Projects\TestSSISSolution\TestSSISProject\ReportAndFail.dtsx	20	1

Figure 5-28. *Results from the SSIS framework metadata query*

The metadata returned from the SSIS framework metadata is more than enough to execute in the SSIS framework execution engine. Our next step is to build the execution engine for the SSIS framework.

Conclusion

In this chapter, we focused on

- Building the metadata database

- Building a test SSIS project

- Adding metadata to the metadata database

The next step is building the execution engine, which we accomplish in the next chapter.

CHAPTER 6

Framework Execution Engine

Now that the metadata database and test application have been built in the previous chapter, it's time to focus on the execution engine. Recall that in the previous chapter, near the beginning, I emphasized the importance of empathy in software architecture and design. I wrote

> *I assume the SSIS developer, or SSIS Developer Team, will be familiar with SSIS, so I build the SSIS framework in SSIS. I want the developer or team to manage and maintain their framework.*

Chapter 5 covered building out the framework and managing the metadata. Now it's time to think about how our SSIS packages will be executed.

Create a Parent SSIS Package

Create a new SSIS project named "SSISFrameworkProject" in a new SSIS solution named "SSISFrameworkSolution." Rename the default SSIS package to "Parent.dtsx" as shown in Figure 6-1.

© Andy Leonard, Kent Bradshaw 2020
A. Leonard and K. Bradshaw, *SQL Server Data Automation Through Frameworks*,
https://doi.org/10.1007/978-1-4842-6213-9_6

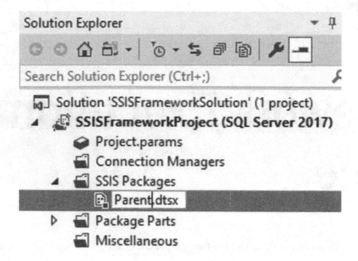

Figure 6-1. *Renaming the Parent.dtsx SSIS package in the SSISFrameworkProject in SSISFrameworkSolution*

Parent.dtsx will

- Log execution values

- Retrieve a list of SSIS application packages from the SSISConfig database – application packages that belong to an SSIS application

- Iterate the retrieved list of application packages

- Log metadata about each application package

- Execute each application package

- Log execution results

Log Execution Values

Begin by adding a String data type package parameter named "ApplicationName," defaulted to "Framework Test," as seen in Figure 6-2.

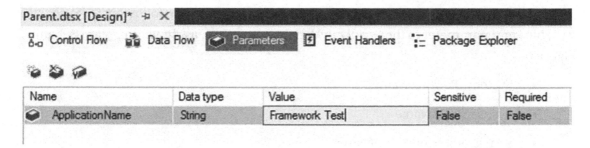

Figure 6-2. *Adding the ApplicationName package parameter to Parent.dtsx*

Leave Sensitive and Required parameter properties set to their default values: False. The ApplicationName parameter contains the name of the SSIS framework SSIS application to be executed when Parent.dtsx is executed. The value of the ApplicationName parameter will be overridden at runtime.

To log execution values, add a Script Task to the Parent.dtsx Control Flow. Rename the Script Task to "SCR Log Initial Values," as shown in Figure 6-3.

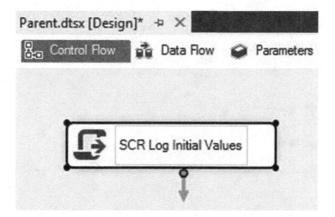

Figure 6-3. *Adding and renaming "SCR Log Initial Values" script task*

The SCR Log Initial Values script task will be used to capture *instrumentation* about the Parent.dtsx package's initial state at execution start. Instrumentation is important in any engineering endeavor. This holds for data engineering, as well as ETL (extract, transform, and load) instrumentation. Capturing the initial state of settings at the start of SSIS package execution is *vital* for troubleshooting. As my friend Grant Fritchey says, "How do you know what's wrong unless you know what *right* looks like?"

Open the SCR Log Initial Values script task editor, and add the following SSIS variables and parameter to the ReadOnlyVariables property, as shown in Figure 6-4 and Figure 6-5.

- System::PackageName

- System::TaskName

- $Package::ApplicationName

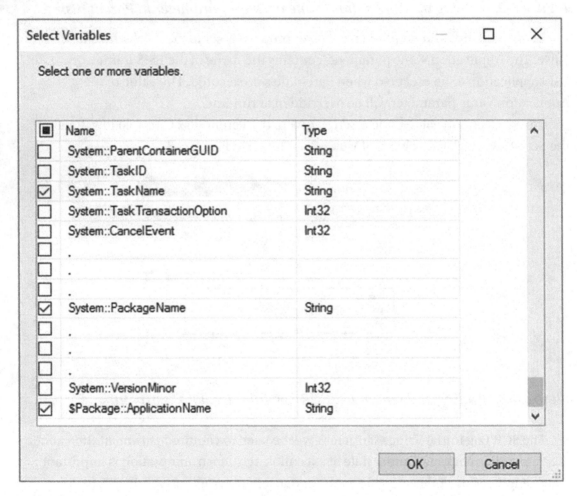

Figure 6-4. *Selecting two SSIS variables and one parameter for the ReadOnlyVariables script task property*

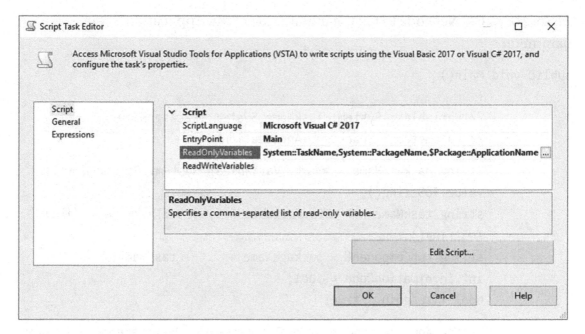

Figure 6-5. *Two SSIS variables and one parameter added to the*
ReadOnlyVariables script task property

Copy the list of variables and parameters from the ReadOnlyVariables script task
property. Click the Edit Script button to open the Visual Studio Tools for Applications
(VSTA) .Net code editor window named VstaProjects. Navigate to `public void Main()`,
and paste the SSIS variables and parameter in a comment, as shown in Figure 6-6.

```
public void Main()
{
    // Variables: System::TaskName,System::PackageName
    // Parameters: $Package::ApplicationName

    Dts.TaskResult = (int)ScriptResults.Success;
}
```

Figure 6-6. *Copying the SSIS variables and parameter into VstaProjects*

Add the .Net C# code shown in Listing 6-1 to log initial values or the
ApplicationName parameter.

Listing 6-1. C# .Net code to log the initial value of the ApplicationName parameter

```
public void Main()
            {
            // Variables: System::TaskName,System::PackageName
            // Parameters: $Package::ApplicationName

            string packageName = ➤ Dts.Variables["System::PackageName"].
            Value.ToString();
            string taskName = ➤ Dts.Variables["System::TaskName"].Value.
            ToString();
            string subComponent = packageName + "." + taskName;
            int informationCode = 1001;
            bool fireAgain = true;

            string applicationName = ➤ Dts.Variables["$Package::Application
            Name"].Value.ToString();
            string description = "ApplicationName: " + applicationName;

            Dts.Events.FireInformation(informationCode, subComponent,➤
            description, "", 0, ref fireAgain);

            Dts.TaskResult = (int)ScriptResults.Success;
            }
```

Once the code in `public void Main()` matches Listing 6-1, the VstaProjects window should configure and raise an SSIS Information event, as shown in Figure 6-7.

```
public void Main()
{
    // Variables: System::TaskName,System::PackageName
    // Parameters: $Package::ApplicationName

    string packageName = Dts.Variables["System::PackageName"].Value.ToString();
    string taskName = Dts.Variables["System::TaskName"].Value.ToString();
    string subComponent = packageName + "." + taskName;
    int informationCode = 1001;
    bool fireAgain = true;

    string applicationName = Dts.Variables["$Package::ApplicationName"].Value.ToString();
    string description = "ApplicationName: " + applicationName;

    Dts.Events.FireInformation(informationCode, subComponent, description, "", 0, ref fireAgain);

    Dts.TaskResult = (int)ScriptResults.Success;
}
```

Figure 6-7. *C# .Net code that raises an SSIS Information event*

Close the VstaProjects window, and click the OK button on the Script Task Editor. Execute the Parent.dtsx SSIS package in the debugger (F5). If all goes well, the SCR Log Initial Values script task should execute and succeed as shown in Figure 6-8.

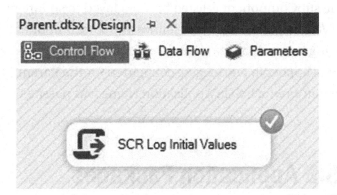

Figure 6-8. *Successful debug execution of Parent.dtsx*

Click the Progress tab (or Execution Results tab if you've already stopped the debugger), and view the Information event coded in the SCR Log Initial Values script task, as shown in Figure 6-9.

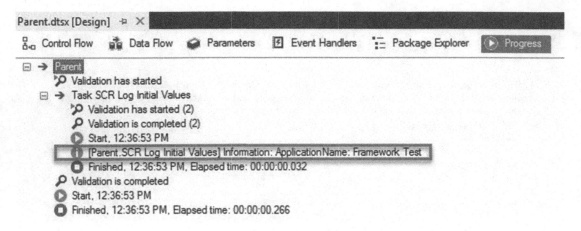

Figure 6-9. *Viewing the Information event*

The text of the Information event is shown in Listing 6-2.

Listing 6-2. Contents of the Information event

`[Parent.SCR Log Initial Values] Information: ApplicationName: Framework Test`

The value of the subComponent is listed first, enclosed in square brackets: `[Parent.`
`SCR Log Initial Values]`. The event type follows the subComponent: `Information`.
The description string is last: `ApplicationName: Framework Test`.

This message will appear in any logging configured for the Parent.dtsx SSIS package
and will inform operators which SSIS Application Name was passed to the Parent.dtsx
package.

Retrieve SSIS Application Packages from SSISConfig

The next step is to retrieve a list of packages to execute in the Parent.dtsx SSIS package.
Add an Execute SQL Task to the Parent.dtsx control flow, and rename it to "SQL Get
Application Packages," as shown in Figure 6-10.

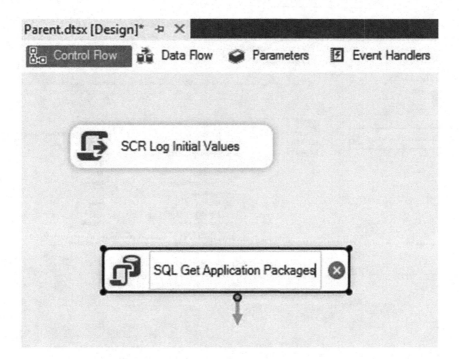

Figure 6-10. *Adding the "SQL Get Application Packages" Execute SQL Task*

Connect a precedence constraint from the SCR Log Initial Values script task to the SQL Get Application Packages execute SQL task, and then open the SQL Get Application Packages execute SQL task editor by right-clicking the task and then clicking "Edit," as shown in Figure 6-11.

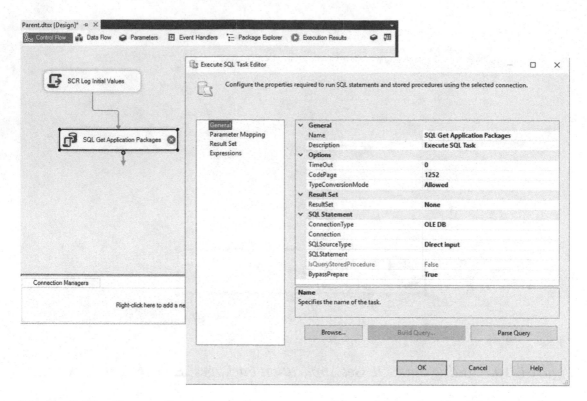

Figure 6-11. *Connecting a precedence constraint and opening the SQL Get Application Packages execute SQL task editor*

Set the SQL Get Application Packages execute SQL task's ConnectionType property to ADO.NET, as shown in Figure 6-12.

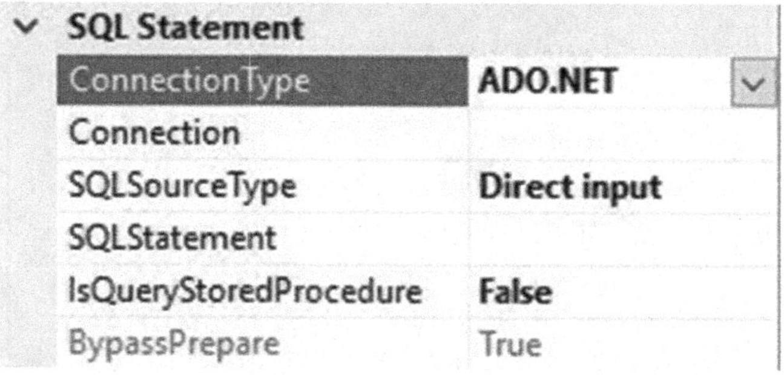

∨ SQL Statement	
ConnectionType	ADO.NET ∨
Connection	
SQLSourceType	Direct input
SQLStatement	
IsQueryStoredProcedure	False
BypassPrepare	True

Figure 6-12. *Setting the ConnectionType to ADO.NET*

Click the drop-down in the Connection property, and click "<New Connection...>" as shown in Figure 6-13.

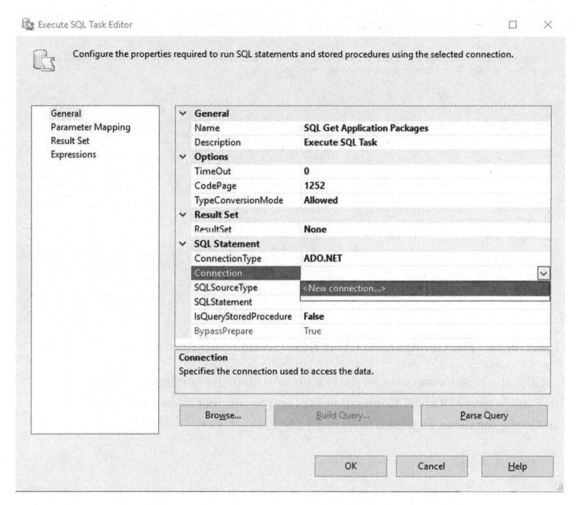

Figure 6-13. *Selecting <New Connection...> from the Connection property*

The Configure ADO.NET Connection Manager dialog displays, as shown in Figure 6-14.

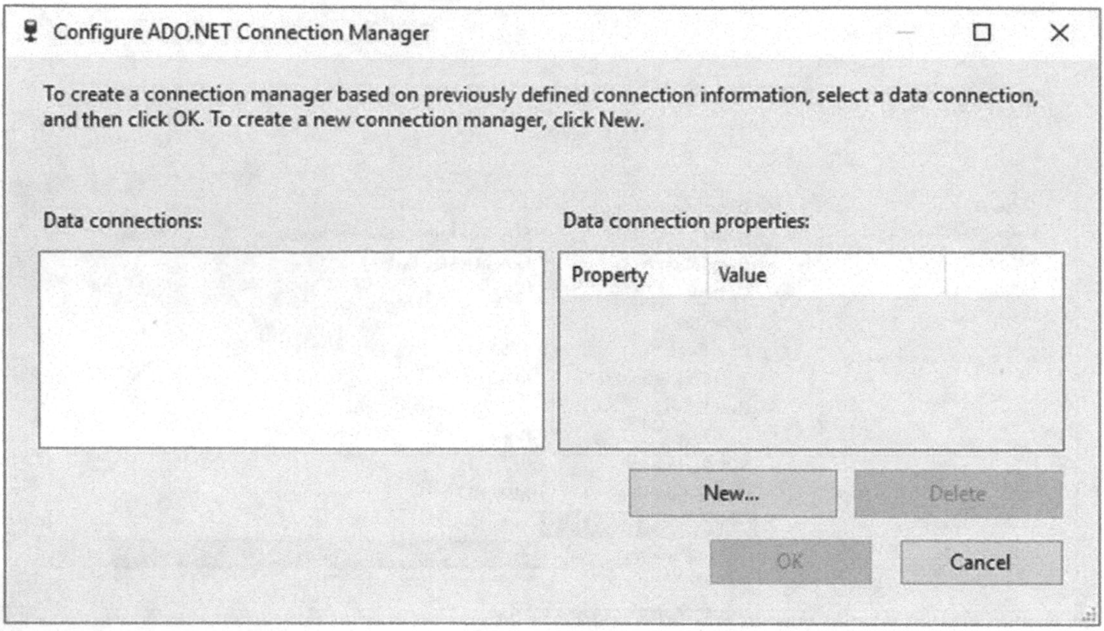

Figure 6-14. *The Configure ADO.NET Connection Manager dialog*

Click the New button to open the Connection Manager configuration dialog. Configure a new ADO.NET Connection Manager to connect to your instance of the SSISConfig database, as shown in Figure 6-15.

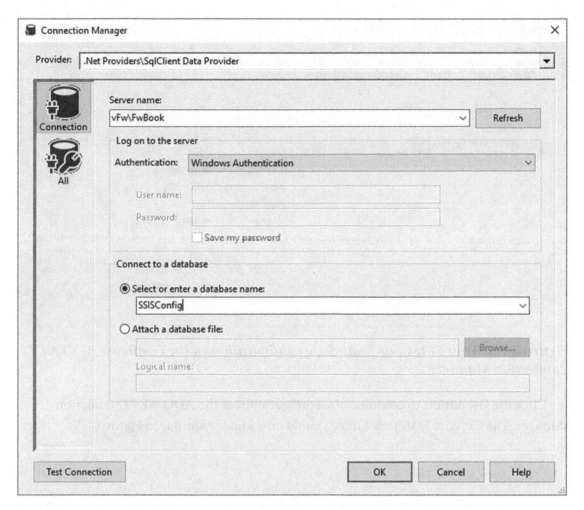

Figure 6-15. *Configuring a new ADO.NET Connection Manager*

Click the OK button to complete configuration of the ADO.NET Connection Manager. The new connection manager configuration contains connection information for your instance of the SSISConfig database and is now stored in the list of configured ADO.NET connection managers as shown in Figure 6-16.

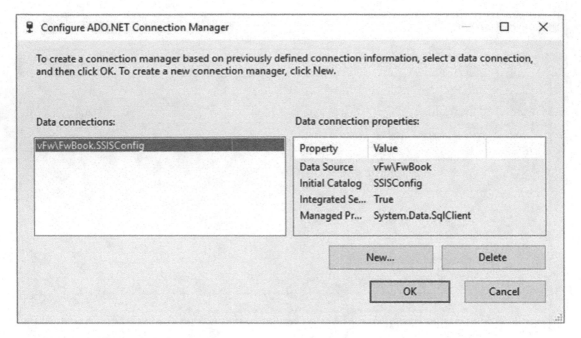

Figure 6-16. *Your SSISConfig database configuration in the Configure ADO.NET Connection Manager*

Click the OK button to complete the configuration of the ADO.NET Connection Manager. The Execute SQL Task Editor should now appear similar to Figure 6-17.

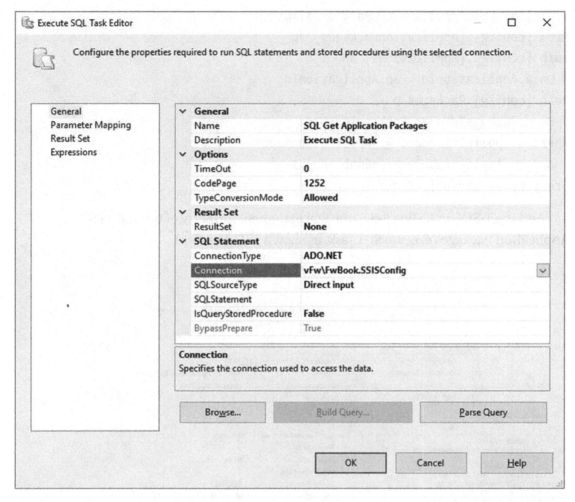

Figure 6-17. *Connection configured for the SQL Get Application Packages execute SQL task*

Listing 5-14 from the previous chapter contains the T-SQL required to return a list of SSIS application packages from the metadata stored in the SSISConfig database for a given SSIS application. Configure the SQL Get Application Packages execute SQL task's SQLStatement property with the T_SQL shown in Listing 6-3.

Listing 6-3. T-SQL to retrieve application packages for an SSIS application from the SSISConfig database

```
Select a.ApplicationName
    , p.PackageLocation + p.PackageName As PackagePath
    , ap.ExecutionOrder
```

```
      , ap.FailApplicationOnPackageFailure
From [config].[ApplicationPackages] ap
Join [config].[Applications] a
  On a.ApplicationId = ap.ApplicationId
Join [config].Packages p
  On p.PackageId = ap.PackageId
Where a.ApplicationName = @ApplicationName
  And ap.ApplicationPackageEnabled = 1
Order By ap.ExecutionOrder
```

Add the T-SQL in Listing 6-3 to the SQLStatement property of the SQL Get Application Packages execute SQL task, as shown in Figure 6-18.

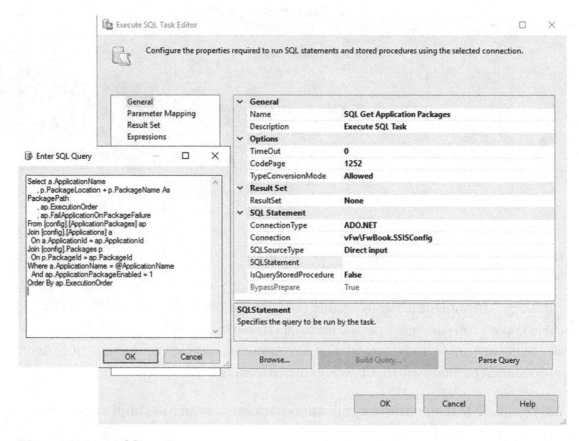

Figure 6-18. *Adding the SQLStatement property's T-SQL*

Click the OK button to close the "Enter SQL Query" dialog. The Execute SQL Task Editor should appear similar to that shown in Figure 6-19.

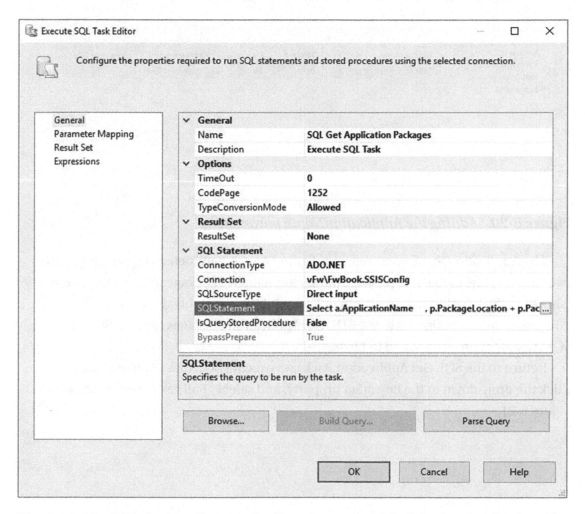

Figure 6-19. *SQL Get Application Packages execute SQL task after configuring the SQLStatement property*

In the list of Execute SQL Task Editor pages on the left side of the Execute SQL Task Editor, click the "Parameter Mapping" page, and then click the Add button to add a parameter mapping. Select the $Package::ApplicationName parameter from the Variable Name column, select the String data type, and enter "ApplicationName" for the Parameter Name column as shown in Figure 6-20.

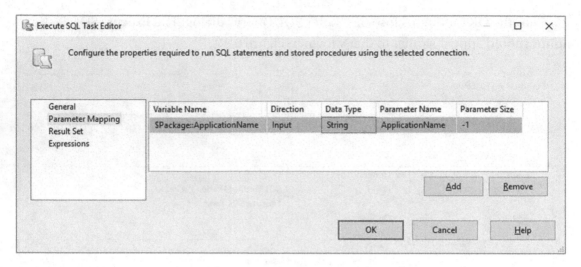

Figure 6-20. *Adding the ApplicationName parameter*

The ApplicationName parameter configured on the Parameter Mapping page will read the value from the SSIS package parameter named $Package::ApplicationName (shown in the Variable Name column in Figure 6-20) and supply the value to the T-SQL query parameter configured in the SQL Get Application Packages execute SQL task's SQLStatement property found in Listing 6-17.

Return to the SQL Get Application Packages execute SQL task's General page. Click the drop-down in the ResultSet property, and select "Full result set," as shown in Figure 6-21.

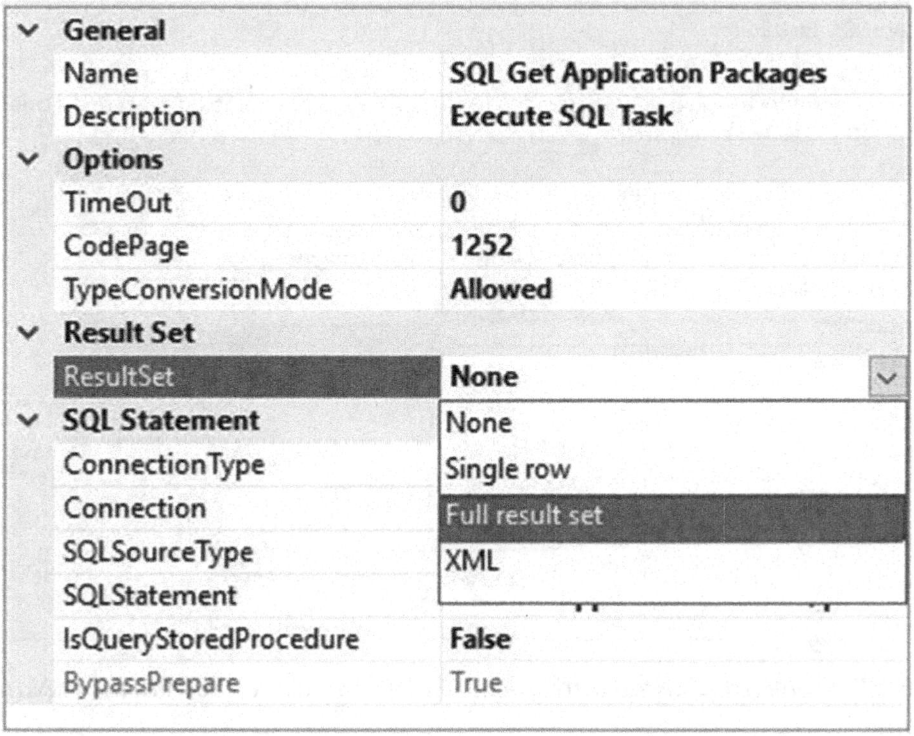

General	
Name	**SQL Get Application Packages**
Description	**Execute SQL Task**
Options	
TimeOut	**0**
CodePage	**1252**
TypeConversionMode	**Allowed**
Result Set	
ResultSet	**None**
SQL Statement	None
ConnectionType	Single row
Connection	Full result set
SQLSourceType	XML
SQLStatement	
IsQueryStoredProcedure	**False**
BypassPrepare	True

Figure 6-21. *Selecting Full result set*

Click the Result Set page, and then click the Add button. Edit the Result Name value, and set it to "0". In the Variable Name drop-down, select "<New variable…>" as shown in Figure 6-22.

115

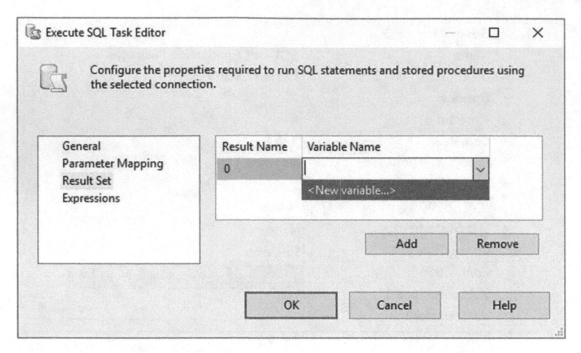

Figure 6-22. *Selecting "<New variable...>" from the Result Set Variable Name drop-down*

When the Add Variable dialog displays, make sure the Container is set to Parent as shown in Figure 6-23.

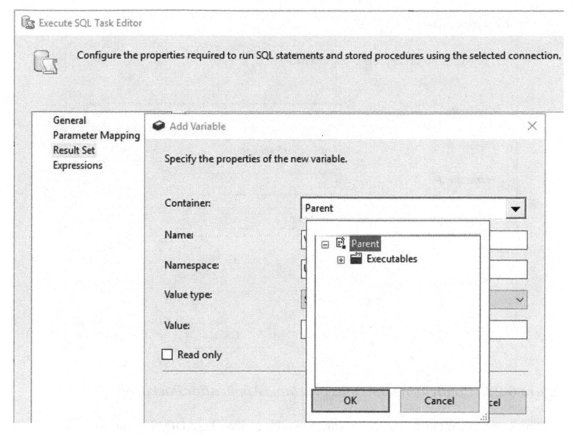

Figure 6-23. *Verifying Parent Container is selected*

Change the Name property to ApplicationPackages, leave the Namespace set to User, and set the Value type to Object as shown in Figure 6-24.

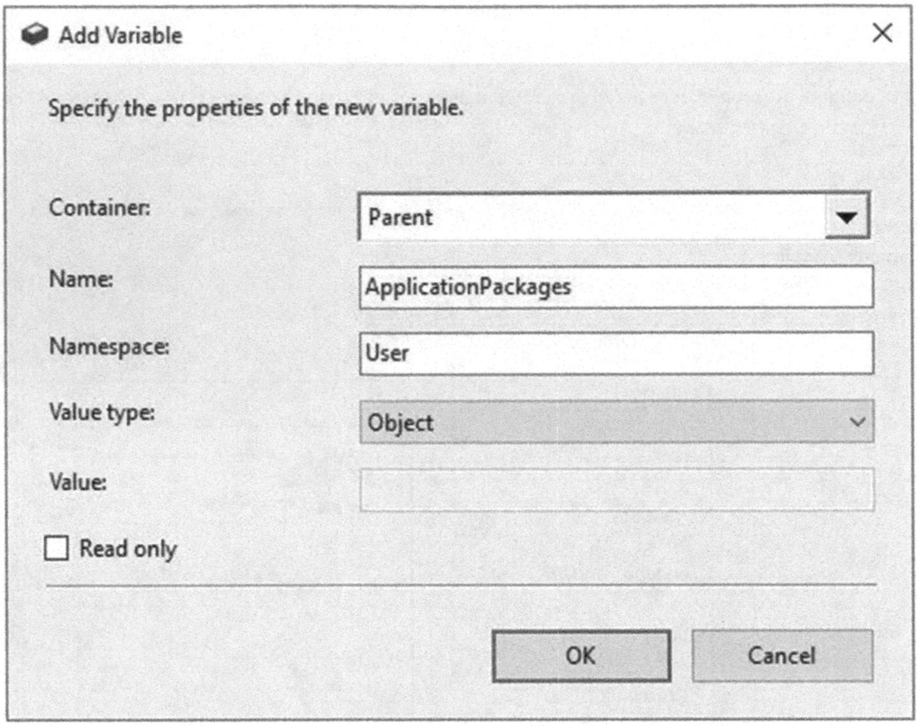

Figure 6-24. *Configuring the Object variable ApplicationPackages*

The User::ApplicationPackages variable will receive the ADO.Net dataset returned from the execution of the T-SQL query stored in the SQL Get Application Packages execute SQL task's SQLStatement property. Since the T-SQL query includes the ApplicationName parameter in the criteria (WHERE clause) and this parameter is set by the $Package::ApplicationName SSIS parameter sent to the Parent.dtsx SSIS package, the ADO.Net dataset will contain the list of packages for which SSISConfig data has been configured. Since the $Package::ApplicationName is defaulted to the SSIS framework application named "Framework Test," the User::ApplicationPackages variable will contain the results shown in Figure 6-25.

	ApplicationName	PackagePath	ExecutionOrder	FailApplicationOnPackageFailure
1	Framework Test	E:\Projects\TestSSISSolution\TestSSISProject\ReportAndSucceed.dtsx	10	1
2	Framework Test	E:\Projects\TestSSISSolution\TestSSISProject\ReportAndFail.dtsx	20	1

Figure 6-25. *Results sent to User::ApplicationPackages variable*

Click the OK button to close the Add Variable dialog.

The General page of the SQL Get Application Packages execute SQL task editor should now appear similar to Figure 6-26.

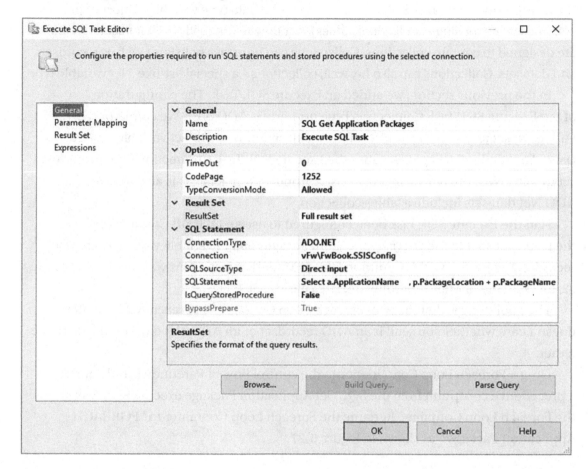

Figure 6-26. *The SQL Get Application Packages execute SQL task's General page*

Click the OK button to close the SQL Get Application Packages execute SQL task editor.

Iterate Application Packages

SSIS Foreach Loop Containers are used to iterate collections of many sorts. The User::ApplicationPackages SSIS variable is an object data type variable. Objects can contain scalars (or single, individual values), but they're not really built for scalars. Objects are designed to contain collections. Collections can be arrays or lists, as well as recordsets and datasets. Collections can also be, well, collections – a special list-like .Net variable type.

In the previous section, we added an Execute SQL Task. The configuration of the Execute SQL Task ConnectionType property is ADO.NET. We configured a Full result set and sent the results (of the result set) into an Object variable named User::ApplicationPackages. Because the Execute SQL Task ConnectionType property is set to ADO.NET, the type of object in User::ApplicationPackages is an ADO.Net dataset. ADO.Net datasets include a tables collection.

Had the Execute SQL Task been configured to use an OLE DB ConnectionType, the result set sent to the User::ApplicationPackages object variable would be an ADO recordset, which is a COM (Common Object Model) object. COM was introduced in the 32-bit (x86) era of Microsoft software, circa early 1990s.

The cool thing is that the SSIS Foreach Loop Container's Foreach ADO Iterator doesn't care whether you send it an ADO recordset or an ADO.Net dataset; it will iterate either.

Drag a Foreach Loop Container onto the control flow of Parent.dtsx and connect a precedence constraint from the SQL Get Application Package execute SQL task to the Foreach Loop Container. Rename the Foreach Loop Container to "FOREACH Application Package" as shown in Figure 6-27.

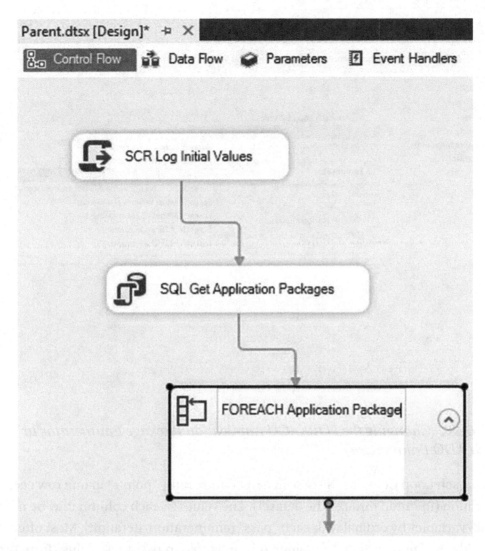

Figure 6-27. *Renaming the Foreach Loop Container*

Open the FOREACH Application Package's editor and navigate to the Collection page. Change the Enumerator to "Foreach ADO Enumerator" as shown in Figure 6-28.

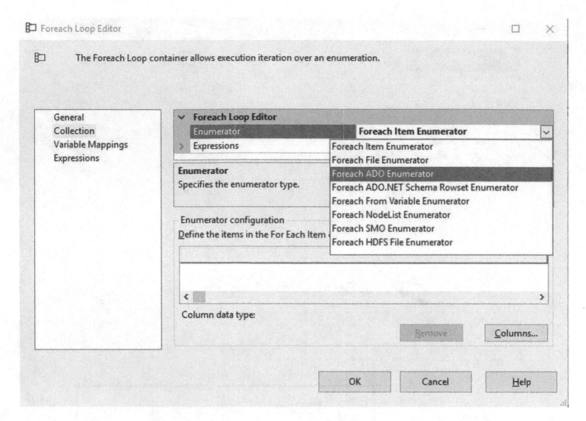

Figure 6-28. *Changing the FOREACH Application Package Enumerator to Foreach ADO Enumerator*

A Foreach Loop Container's Foreach ADO Enumerator "points" to one row each enumeration (iteration through the dataset). The values in each column may be mapped into SSIS variables by ordinal with each "pass" (enumeration/iteration). Most often, SSIS tasks inside the Foreach Loop Container are configured to *read* these values from the variables and act upon them in some way. More on Variable Mappings in a bit.

Each enumerator presents an enumerator-specific properties view. In the Enumerator configuration group box, click the "ADO object source variable" drop-down, and select the User::ApplicationPackages variable as shown in Figure 6-29.

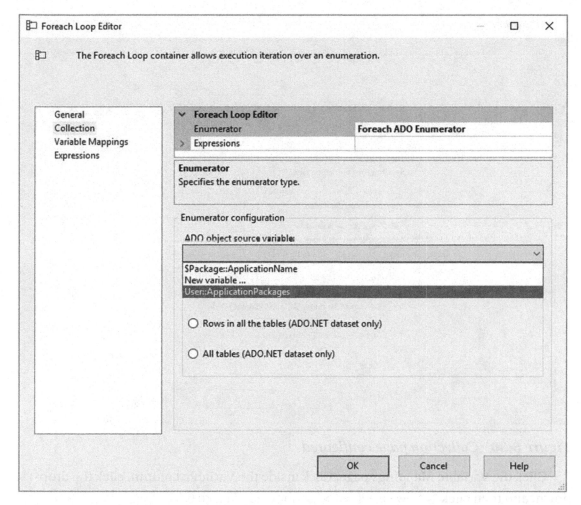

Figure 6-29. *Selecting the ADO object source variable*

Leave the Enumeration mode set to "Rows in the first table." The Collection page is configured as shown in Figure 6-30.

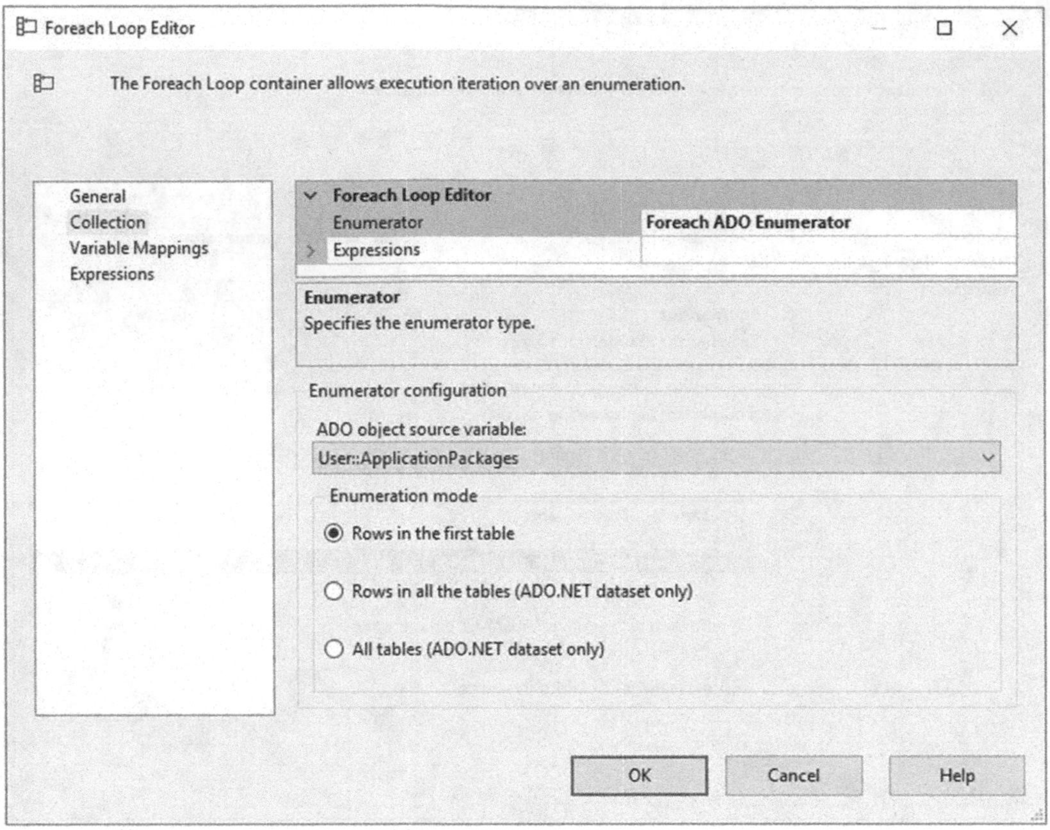

Figure 6-30. *Collection page configured*

Click the Variable Mappings page. Click inside the Variable column, click the drop-down, and then click "<New variable...>" as shown in Figure 6-31.

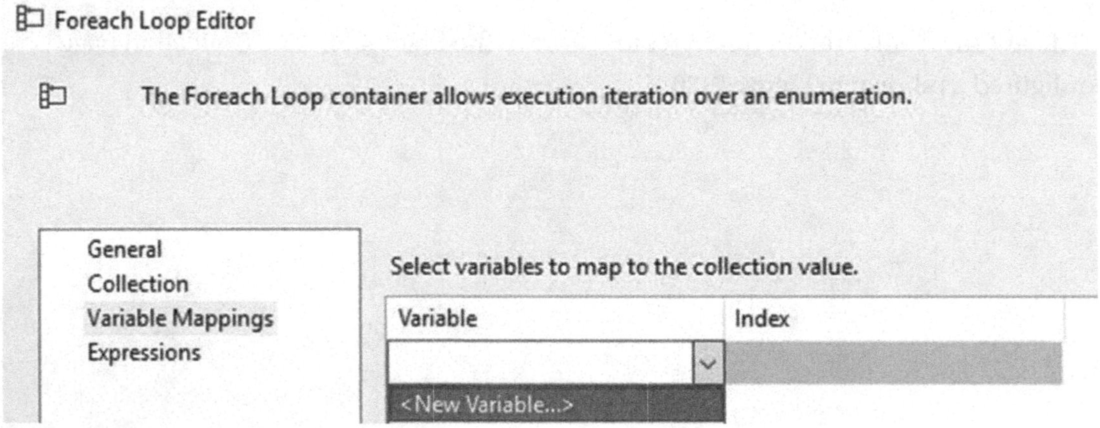

Figure 6-31. *Clicking New Variable on the Variable Mappings page*

As before, make sure the Container is set to the SSIS package scope ("Parent") when the Add Variable dialog displays. Enter "PackagePath" as the variable name, and leave all other settings at their respective defaults, as shown in Figure 6-32.

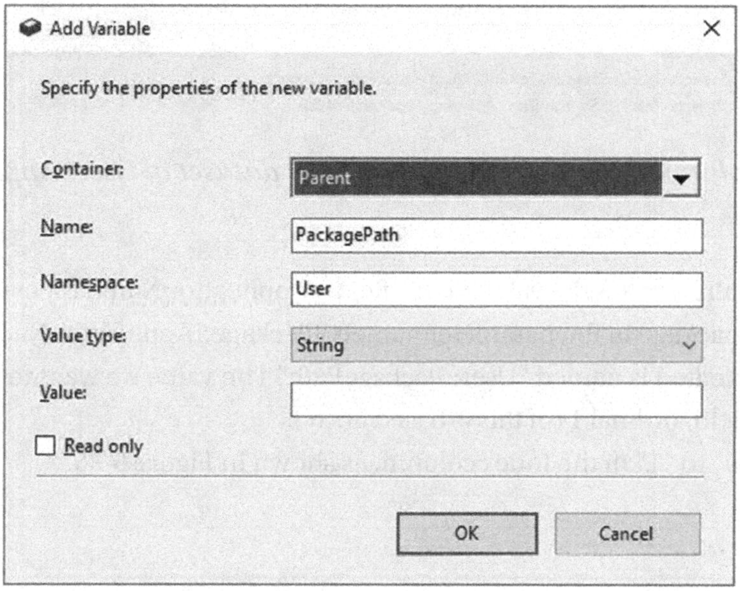

Figure 6-32.   Adding the PackagePath variable

Click the OK button to close the Add Variable dialog. Note the Index column defaults to "0" for this first variable assignment as shown in Figure 6-33.

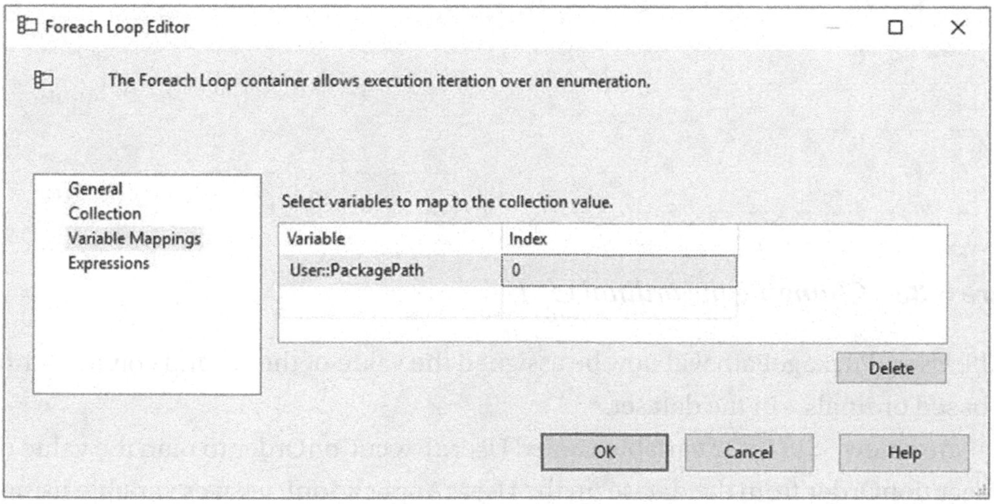

Figure 6-33.   User::PackagePath assigned to ordinal "0"

What does "0" mean? I'm glad you asked. "0" indicates the ordinal of the column in the User::ApplicationPackages dataset. Please recall the contents of Tables(0) in the ADO.Net dataset as shown in Figure 6-34.

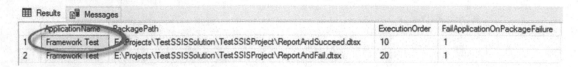

	ApplicationName	PackagePath	ExecutionOrder	FailApplicationOnPackageFailure
1	Framework Test	E:\Projects\TestSSISSolution\TestSSISProject\ReportAndSucceed.dtsx	10	1
2	Framework Test	E:\Projects\TestSSISSolution\TestSSISProject\ReportAndFail.dtsx	20	1

Figure 6-34. *Column Ordinal "0" in the ADO.Net dataset in User::ApplicationPackages*

Our package already *has* the value in this field – ApplicationName is passed to the Parent.dtsx SSIS package in the parameter named $Package::ApplicationName. The SSIS variable we just created is named "User::PackagePath". The value we want to assign here is in the second field, ordinal 1 (of this 0-based array).

Change the "0" to "1" in the Index column, as shown in Figure 6-35.

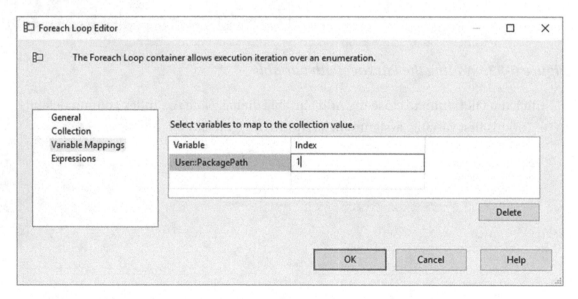

Figure 6-35. *Changing the ordinal to "1"*

The User::PackagePath will now be assigned the value of the second column – 1 in zero-based ordinals – in the dataset.

Create a new SSIS Int32 variable named User::ExecutionOrder to map the value of the ExecutionOrder from the dataset in the User::ApplicationPackages variable using the following settings on the Add Variable dialog, as shown in Figure 6-36.

- Container: Parent

- Name: ExecutionOrder

- Namespace: User

- Value type: Int32

- Value: -1

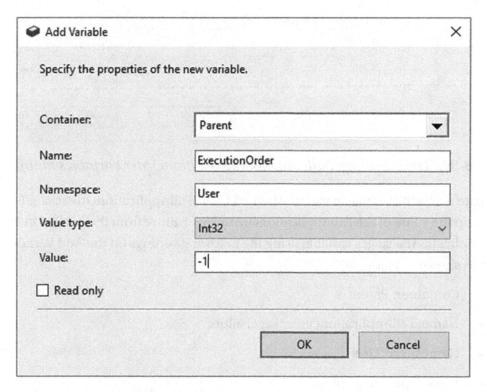

Figure 6-36. *Mapping the ExecutionOrder SSIS variable*

Update the Index column to 2 for the ExecutionOrder column. The FOREACH Application Package Foreach Loop Container's Variable Mappings page should appear similar to Figure 6-37.

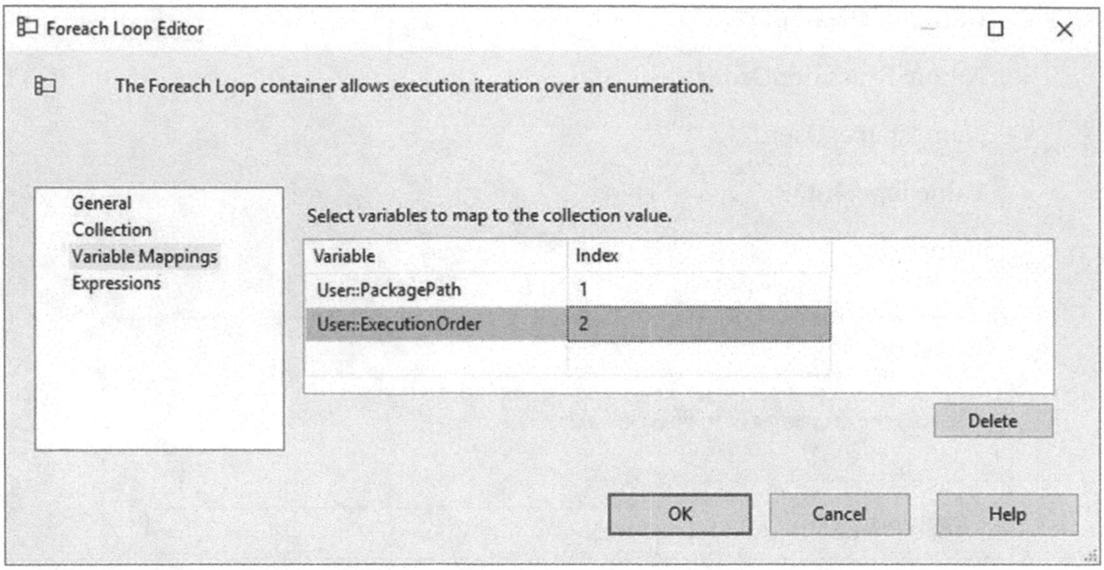

Figure 6-37. *User::PackagePath and User::ExecutionOrder variables mapped*

Create a new SSIS Boolean variable named User::FailApplicationOnPackageFail
ure to map the value of the FailApplicationOnPackageFailure from the dataset in the
User::ApplicationPackages variable using the following settings on the Add Variable
dialog, as shown in Figure 6-38.

- Container: Parent

- Name: FailApplicationOnPackageFailure

- Namespace: User

- Value type: Boolean

- Value: true

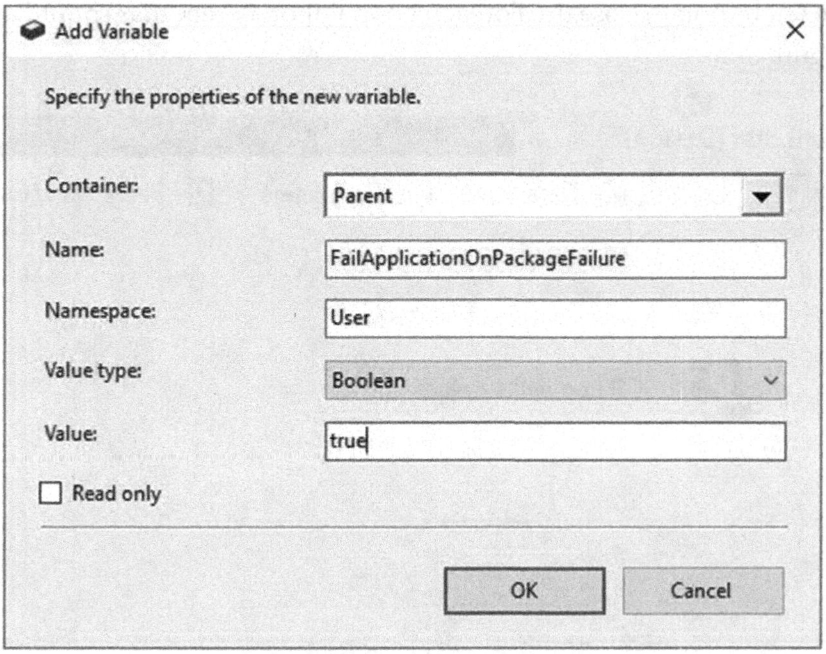

Figure 6-38. *Mapping the FailApplicationOnPackageFailure SSIS variable*

Update the Index column to 3 for the FailApplicationOnPackageFailure column. The FOREACH Application Package Foreach Loop Container's Variable Mappings page should appear similar to Figure 6-39.

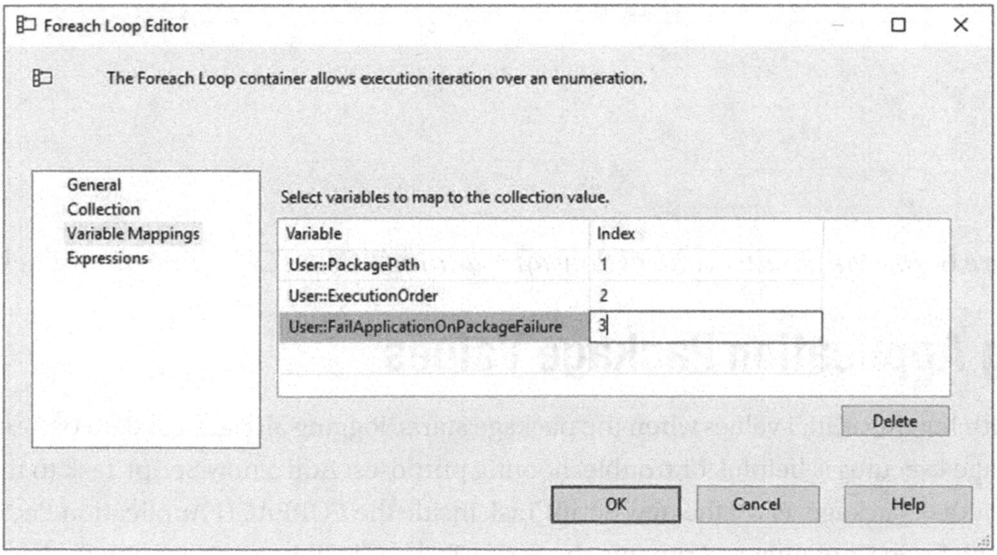

Figure 6-39. *User:: FailApplicationOnPackageFailure variable mapped*

Click the OK button to close the Foreach Loop Editor. Parent.dtsx should appear similar to Figure 6-40.

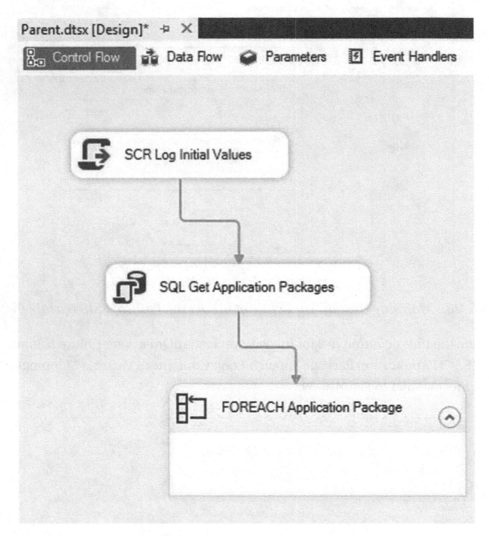

Figure 6-40. *Parent.dtsx control flow after adding FOREACH Application Package*

Log Application Package Values

As with logging initial values when the package starts, logging all package data before the package executes is helpful for troubleshooting purposes. Add a new Script Task to the Parent.dtsx package. Place the new Script Task inside the FOREACH Application Package Foreach Loop Container and rename the Script Task to "SCR Log Application Package Values" as shown in Figure 6-41.

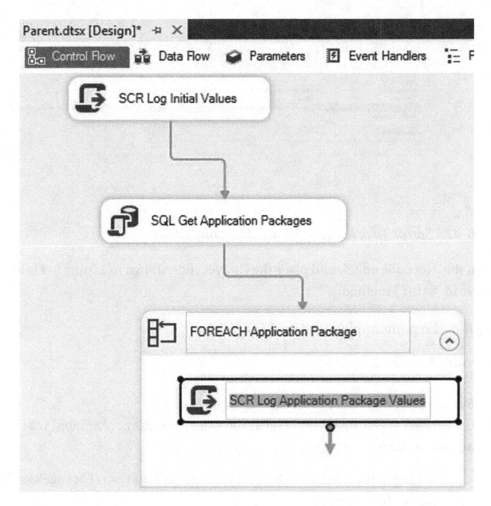

Figure 6-41. *Adding the SCR Log Application Package Values Script Task*

Open the Script Task Editor and add the following SSIS variables to the ReadOnlyVariables property:

- System::PackageName

- System::TaskName

- User::PackagePath

- User::ExecutionOrder

- User::FailApplicationOnPackageFailure

The Script Task's Script page should appear as shown in Figure 6-42.

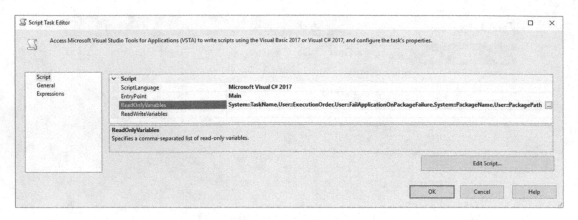

Figure 6-42. *Script Task ReadOnlyVariables configured*

Open the .Net code editor, and place the C# .Net code shown in Listing 6-4 into the `public void Main()` method.

Listing 6-4. Logging Application Package Values

```
public void Main()
  {
    // System::PackageName, System::TaskName
    // User::PackagePath, User::ExecutionOrder, ➤ User::FailApplicationOnP
       ackageFailure

            string packageName = ➤ Dts.Variables["System::PackageName"].
            Value.ToString();
            string taskName = ➤ Dts.Variables["System::TaskName"].Value.
            ToString();
            string subComponent = packageName + "." + taskName;
            int informationCode = 1001;
            bool fireAgain = true;

            string packagePath = ➤ Dts.Variables["User::PackagePath"].
            Value.ToString();
            string description = "PackagePath: " + packagePath;
            Dts.Events.FireInformation(informationCode, subComponent, ➤
            description, "", 0, ref fireAgain);
```

```
int executionOrder = ➤ Convert.ToInt32(Dts.Variables["User::Ex
ecutionOrder"].Value);
description = "ExecutionOrder: " + executionOrder.ToString();
Dts.Events.FireInformation(informationCode, subComponent, ➤
description, "", 0, ref fireAgain);

bool failApplicationOnPackageFailure = ➤ Convert.
ToBoolean(Dts.Variables["User::FailApplicationOnPackageFailu
re"]➤
```
.Value);
```
description = "FailApplicationOnPackageFailure: " + ➤
failApplicationOnPackageFailure.ToString();
Dts.Events.FireInformation(informationCode, subComponent, ➤
description, "", 0, ref fireAgain);

Dts.TaskResult = (int)ScriptResults.Success;
}
```

When the C# .Net code is added to the VstaProjects editor for the SCR Log Application Package Values script task, it should appear as shown in Figure 6-43.

```
public void Main()
{
    // System::PackageName, System::TaskName
    // User::PackagePath, User::ExecutionOrder, User::FailApplicationOnPackageFailure

    string packageName = Dts.Variables["System::PackageName"].Value.ToString();
    string taskName = Dts.Variables["System::TaskName"].Value.ToString();
    string subComponent = packageName + "." + taskName;
    int informationCode = 1001;
    bool fireAgain = true;

    string packagePath = Dts.Variables["User::PackagePath"].Value.ToString();
    string description = "PackagePath: " + packagePath;
    Dts.Events.FireInformation(informationCode, subComponent, description, "", 0, ref fireAgain);

    int executionOrder = Convert.ToInt32(Dts.Variables["User::ExecutionOrder"].Value);
    description = "ExecutionOrder: " + executionOrder.ToString();
    Dts.Events.FireInformation(informationCode, subComponent, description, "", 0, ref fireAgain);

    bool failApplicationOnPackageFailure = Convert.ToBoolean(Dts.Variables["User::FailApplicationOnPackageFailure"].Value);
    description = "FailApplicationOnPackageFailure: " + failApplicationOnPackageFailure.ToString();
    Dts.Events.FireInformation(informationCode, subComponent, description, "", 0, ref fireAgain);

    Dts.TaskResult = (int)ScriptResults.Success;
}
```

Figure 6-43. *SCR Log Application Package Values C# .Net code*

Close the VstaProjects window and click the OK button to close the Script Task Editor. The control flow for Parent.dtsx should appear as shown in Figure 6-44.

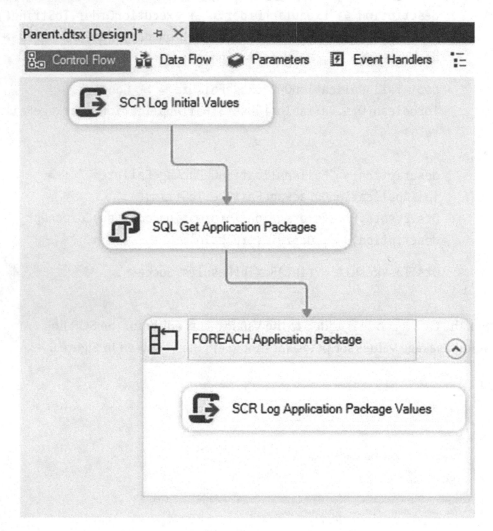

Figure 6-44. *Parent.dtsx control flow after adding SCR Log Application Package Values*

Test the new functionality by executing the Parent.dtsx package in the debugger. Press the F5 key, and then observe the Progress tab once debug execution completes, as shown in Figure 6-45.

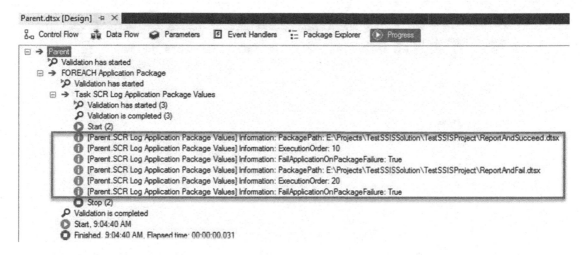

Figure 6-45. *Viewing instrumentation messages raised by SCR Log Application Package Values*

Instrumentation is nearly complete for the Parent.dtsx package, the SSIS framework execution engine. The next step is to execute the child packages for which metadata has been returned from the SSIS framework database.

Execute Application Packages

There are a number of mechanisms available for executing SSIS packages, including

- .Net (C# and Visual Basic)

- PowerShell

- DtExec command-line

- SQL Agent

- SSIS Execute Package Task

In this example, we will use the SSIS Execute Package Task. Drag an Execute Package Task into the FOREACH Application Package Foreach Loop Container. Connect a precedence constraint to the Execute Package Task from SCR Log Application Package Values, and then rename the Execute Package Task to "EPT Execute Child Package," as shown in Figure 6-46.

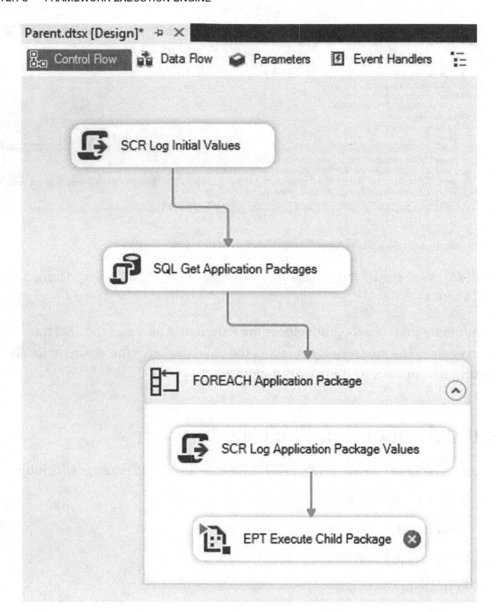

Figure 6-46. *Adding the EPT Execute Child Package Execute Package Task*

Open the Execute Package Task Editor and navigate to the Package page. Change the ReferenceType property to External Reference, as shown in Figure 6-47.

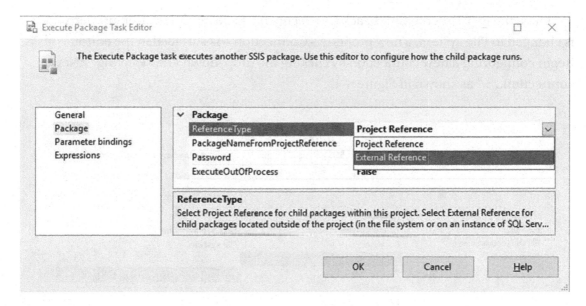

Figure 6-47. *Changing the Execute Package Task's ReferenceType property to External Reference*

Change the EPT Execute Child Package execute package task's Location property to "File system," as shown in Figure 6-48.

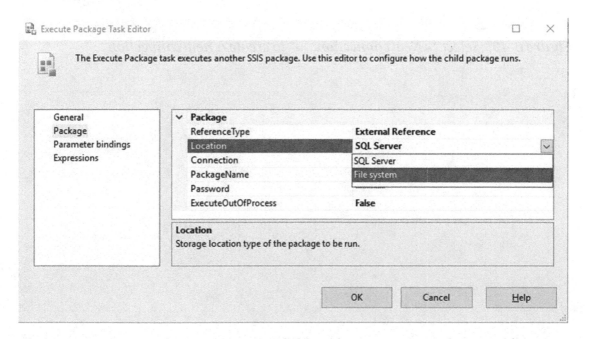

Figure 6-48. *Setting the EPT Execute Child Package execute package task's Location property*

Once the EPT Execute Child Package execute package task's Location property is changed to File system, a new property – Connection – is surfaced in the editor. Begin configuring a new connection by clicking the drop-down and selecting "<New connection...>" as shown in Figure 6-49.

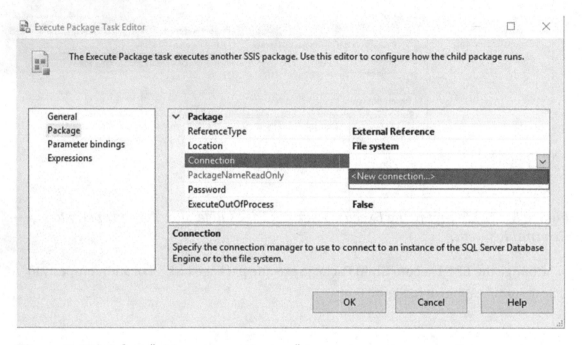

Figure 6-49. *Select "<New connection...>" to create a new connection*

Because the ReferenceType property is set to External Reference and the Location property is set to File system, the EPT Execute Child Package execute package task's Connection property expects to connect to a file – an *SSIS Package file* or dtsx file – so the New Connection operation creates a new File Connection Manager and opens its editor, as shown in Figure 6-50.

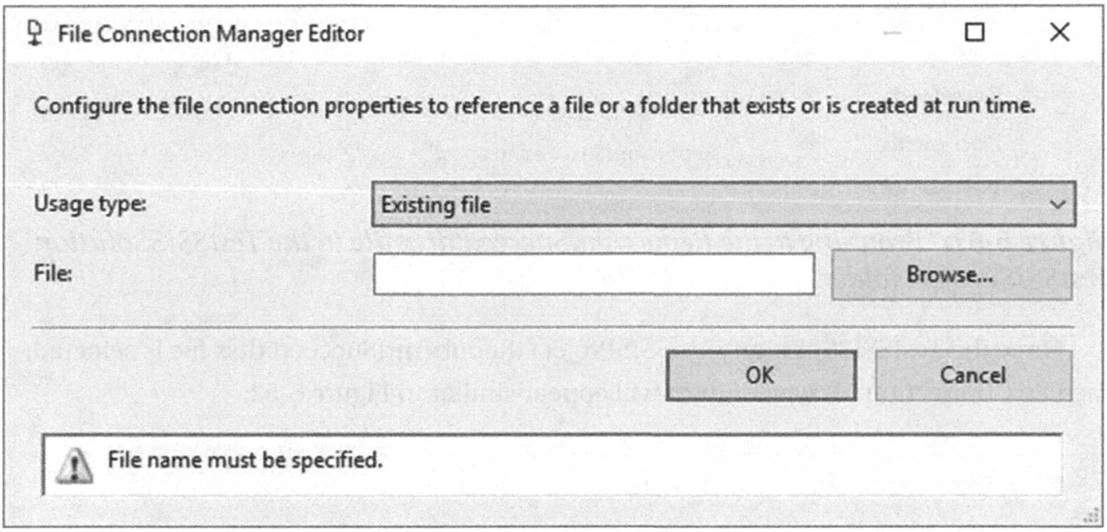

Figure 6-50. *A File Connection Manager ready for configuration*

Leave the "Usage type" file connection manager property set to "Existing file." Click the Browse button and navigate to the ReportAndSucceed.dtsx file in the TestSSISSolution\TestSSISProject folder, as shown in Figure 6-51.

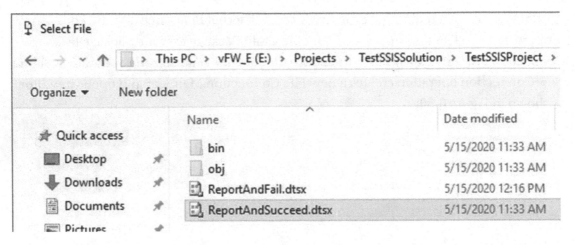

Figure 6-51. *Browsing to the ReportAndSucceed.dtsx file in the TestSSISSolution*
TestSSISProject folder

Once the TestSSISSolution\TestSSISProject\ReportAndSucceed.dtsx file is selected,
the File Connection Manager Editor will appear similar to Figure 6-52.

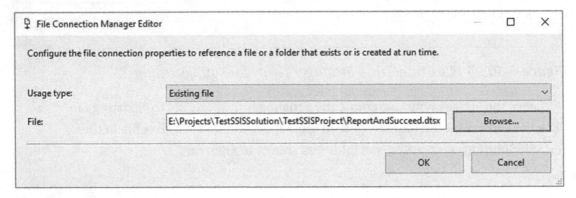

Figure 6-52. *TestSSISSolution\TestSSISProject\ReportAndSucceed.dtsx file*
selected

Click the OK button on the File Connection Manager Editor to return to the EPT
Execute Child Package execute package task editor, which appear similar to Figure 6-53.

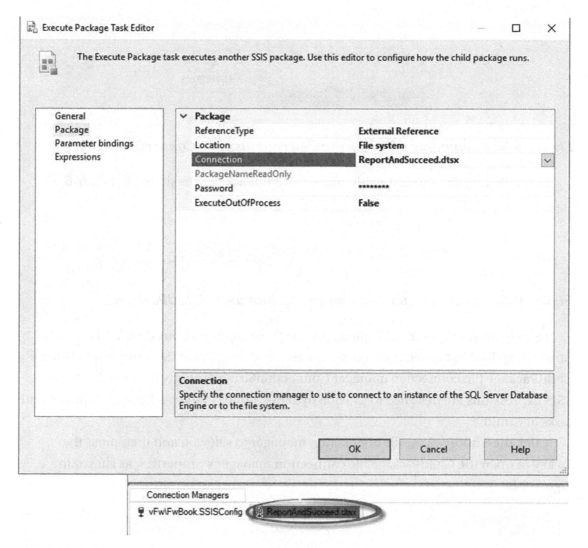

Figure 6-53. *Configured EPT Execute Child Package and ReportAndSucceed.dtsx File Connection Manager*

Click the OK button on the EPT Execute Child Package execute package task editor.

Right-click the ReportAndSucceed.dtsx file connection manager, and then click Rename as shown in Figure 6-54.

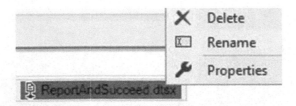

Figure 6-54. *Renaming the ReportAndSucceed.dtsx File Connection Manager*

Rename the File Connection Manager to "ChildPackage," as shown in Figure 6-55.

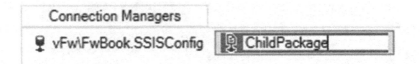

Figure 6-55. *Renaming the File Connection Manager "ChildPackage"*

In order to execute the SSIS application package supplied from FOREACH Application Package (which gets metadata from querying the SSISConfig database), the ChildPackage file connection manager ConnectionString property needs to be dynamic. SSIS Expressions are designed to provide dynamic values to the package, containers, and tasks at runtime.

Click the ChildPackage file connection manager to select it, and then press the F4 key to open the ChildPackage file connection manager's properties, as shown in Figure 6-56.

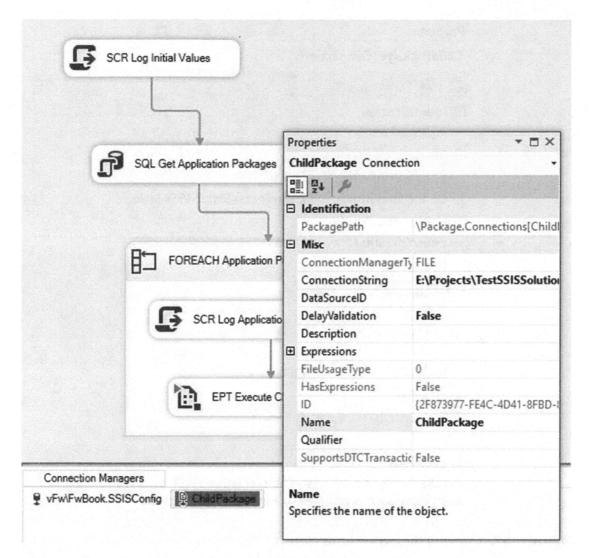

Figure 6-56. *Opening Properties for the ChildPackage file connection manager*

Click the Expressions property, and then click the ellipsis in the Expressions property value area, as shown in Figure 6-57.

Figure 6-57. *Clicking the Expressions ellipsis*

Clicking the Expressions property ellipsis opens the Property Expressions Editor, as shown in Figure 6-58.

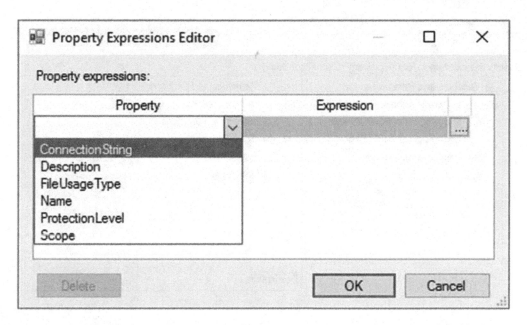

Figure 6-58. *File Connection Manager Property Expressions Editor dialog properties*

Select the ConnectionString property from the Property drop-down as shown in Figure 6-59.

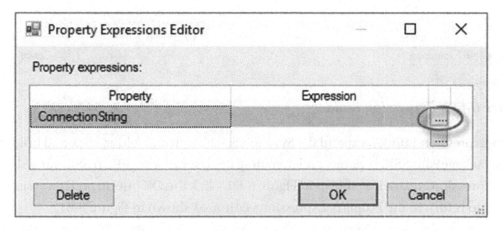

Figure 6-59. *Selecting the ConnectionString property of the File Connection Manager*

Click the ellipsis in the Expression column of the Property Expressions Editor (circled in Figure 6-59) to open the Expression Builder, as shown in Figure 6-60.

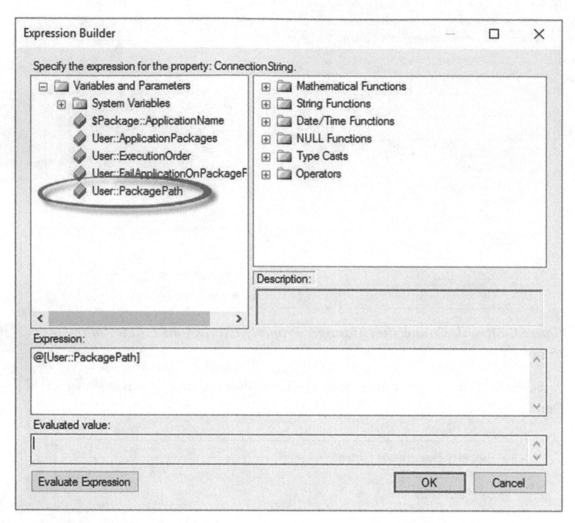

Figure 6-60. *Expression Builder*

In Expression Builder, expand the System Variables virtual folder, click and hold the User::PackagePath SSIS variable, and then drag the User::PackagePath SSIS variable into the "Expression" textbox as shown in Figure 6-60. Click the OK button on Expression Builder to return to the Property Expressions Editor, as shown in Figure 6-61.

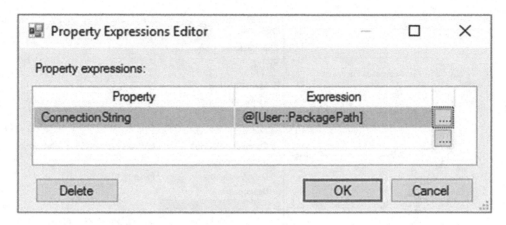

Figure 6-61. *Property Expressions Editor with a configured ConnectionString property*

Figure 6-61 displays the ChildPackage file connection manager's ConnectionString property that is now dynamic and driven by the value of the User::PackagePath SSIS variable. Remember, the User::PackagePath SSIS variable value is managed in the FOREACH Application Package foreach loop container, where the User::PackagePath SSIS variable is populated by each PackagePath value in the ADO.Net dataset returned from the query for metadata stored in the SSISConfig database. Thus, Parent.dtsx, as now configured (once the OK button is clicked), will execute *each and every* (enabled) SSIS framework application package associated with the SSIS application.

Click the OK button to close the Property Expressions Editor. Expand the Expressions property (collection) for the ChildPackage file connection manager, and note the ConnectionString property is now managed by the value of the User::PackagePath SSIS variable. Note also the ChildPackage file connection manager contains Expressions as indicated by the f(x) decoration, as shown in Figure 6-62.

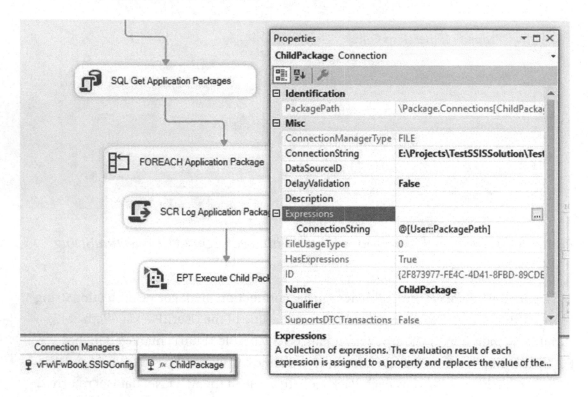

Figure 6-62. *ConnectionString Override and Expressions Decoration for ChildPackage file connection manager*

Test the operation of Parent.dtsx by starting the SSIS debugger (F5). At first glance, the results may appear askew, as shown in Figure 6-63.

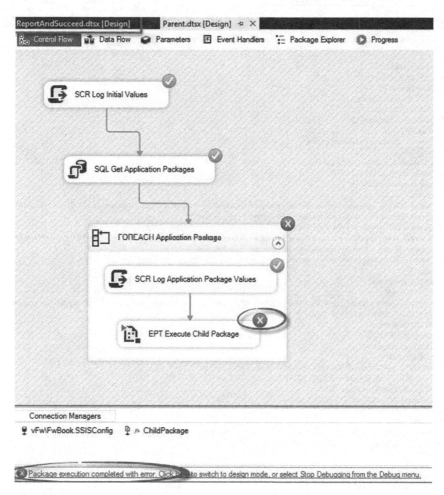

Figure 6-63. *Test Execution Results*

Note the test execution of Parent.dtsx failed. The lower package execution message affirms package execution failure, as does task-failed indications on the EPT Execute Child Package execute package task and the FOREACH Application Package foreach loop container.

What happened? Great question! Open the Progress/Execution Results tab on the Parent.dtsx package tab and view the messages recorded during execution, as shown in Figure 6-64.

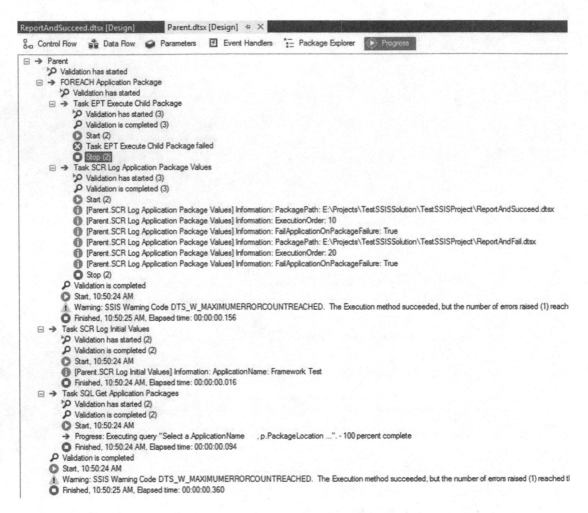

Figure 6-64. *Viewing the Progress/Execution Results tab*

Figure 6-64 shows the messages raised during the failed execution of Parent.dtsx. We see metadata from both files in the SSIS application – ReportAndSucceed.dtsx and ReportAndFail.dtsx – logged by the SCR Log Application Package Values script task. What's missing is an indication of whether each SSIS package execution succeeded or failed.

Log Execution Results

To begin logging execution results, drag a new Script Task into the FOREACH Application Package foreach loop container, and then rename the new Script Task to "SCR Log Package Execution Success," as shown in Figure 6-65.

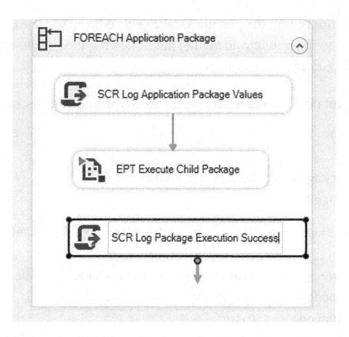

Figure 6-65. *Adding the SCR Log Package Execution Success script task*

Open the SCR Log Package Execution Success script task editor. As before, add System::PackageName, System::TaskName, and User::PackagePath SSIS variables to the ReadOnlyVariables property, as shown in Figure 6-66.

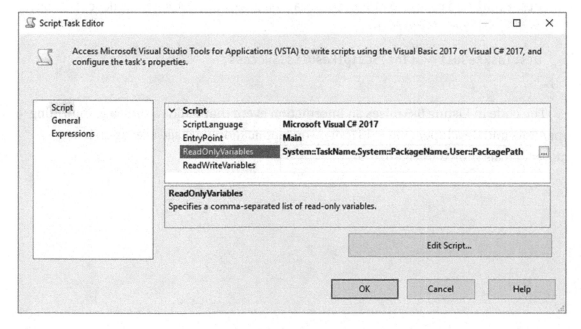

Figure 6-66. *Populating the ReadOnlyVariables property*

Click the Edit Script button to open the VstaProjects editor. Add the C# .Net code shown in Listing 6-5 to the public void Main() method to log the successful execution of an SSIS application package in the framework.

Listing 6-5. C# .Net code to log a successful SSIS application package execution

```csharp
public void Main()
  {
    // System::PackageName, System::TaskName
    // User::PackagePath

    string packageName = ➤ Dts.Variables["System::PackageName"].Value.
    ToString();
    string taskName = Dts.Variables["System::TaskName"].Value.ToString();
    string subComponent = packageName + "." + taskName;
    int informationCode = 1001;
    bool fireAgain = true;

    string packagePath = Dts.Variables["User::PackagePath"].Value.
    ToString();
    string description = packagePath + " execution succeeded";
    Dts.Events.FireInformation(informationCode, subComponent, description,
    ➤ "", 0, ref fireAgain);

    Dts.TaskResult = (int)ScriptResults.Success;
  }
```

The code in Listing 6-5 raises an Information event that builds a message informing operators and developers the SSIS framework application package execution has succeeded, as shown in Figure 6-67.

```
public void Main()
{
    // System::PackageName, System::TaskName
    // User::PackagePath

    string packageName = Dts.Variables["System::PackageName"].Value.ToString();
    string taskName = Dts.Variables["System::TaskName"].Value.ToString();
    string subComponent = packageName + "." + taskName;
    int informationCode = 1001;
    bool fireAgain = true;

    string packagePath = Dts.Variables["User::PackagePath"].Value.ToString();
    string description = packagePath + " execution succeeded";
    Dts.Events.FireInformation(informationCode, subComponent, description, "", 0, ref fireAgain);

    Dts.TaskResult = (int)ScriptResults.Success;
}
```

Figure 6-67. *Raising an Information event on successful SSIS framework application package execution*

Close the VstaProjects window and then click the OK button on the SCR Log Package Execution Success script task editor. Connect an on-success precedence constraint from the EPT Execute Child Package execute package task to the SCR Log Package Execution Success script task, as shown in Figure 6-68.

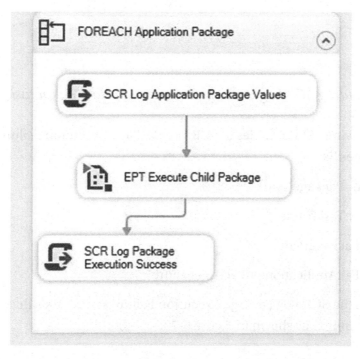

Figure 6-68. *SCR Log Package Execution Success configured*

Drag another Script Task into the FOREACH Application Package foreach loop container, and rename the Script Task to "SCR Log Package Execution Failure," and connect an on-failure precedence constraint (red arrow) from the EPT Execute Child Package execute package task to the SCR Log Package Execution Failure script task, as shown in Figure 6-69.

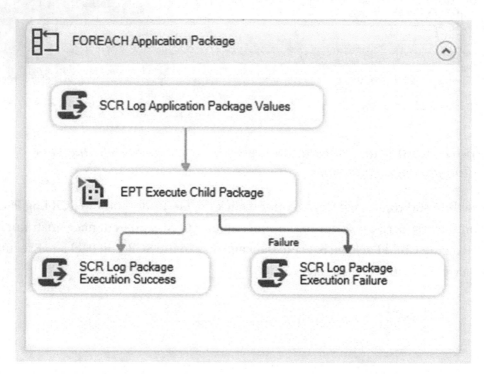

Figure 6-69. *Adding SCR Log Package Execution Failure script task*

Add the following SSIS variables to SCR Log Package Execution Failure script task's ReadOnlyVariables list:

- System::PackageName

- System::TaskName

- User::PackagePath

- User::FailApplicationOnPackageFailure

Once added, the SCR Log Package Execution Failure script task's ReadOnlyVariables property should appear as shown in Figure 6-70.

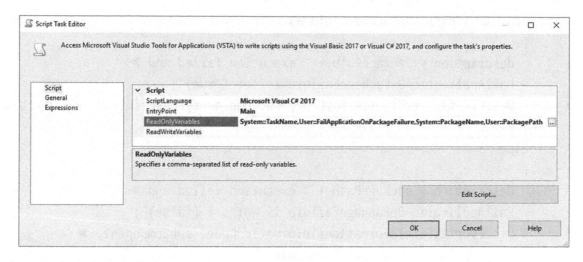

Figure 6-70. *Viewing the SCR Log Package Execution Failure script task's ReadOnlyVariables property*

Click the Edit Script button and add the C# .Net code shown in Listing 6-6 to the `public void Main()` method.

Listing 6-6. C# .Net code to log application package execution failure

```csharp
public void Main()
  {
    // System::PackageName, System::TaskName
    // User::FailApplicationOnPackageFailure, User::PackagePath

    string packageName = ➤ Dts.Variables["System::PackageName"].Value.
    ToString();
    string taskName = Dts.Variables["System::TaskName"].Value.ToString();
    string subComponent = packageName + "." + taskName;
    int informationCode = 1001;
    int errorCode = -999;
    bool fireAgain = true;

    bool failApplicationOnPackageFailure = ➤ Convert.ToBoolean(Dts.Variables
    ["User::FailApplicationOnPackageFailure"] ➤
.Value);
    string packagePath = Dts.Variables["User::PackagePath"].Value.ToString();
    string description = String.Empty;
```

```
    if(failApplicationOnPackageFailure)
      {
        description = packagePath + " execution failed and ➤
        FailApplicationOnPackageFailure is set (true)";
        Dts.Events.FireError(errorCode, subComponent, description, "", 0);
      }
    else
      {
        description = packagePath + " execution failed and ➤
        FailApplicationOnPackageFailure is NOT set (false)";
        Dts.Events.FireInformation(informationCode, subComponent, ➤
        description, "", 0, ref fireAgain);
      }

    Dts.TaskResult = (int)ScriptResults.Success;
  }
```

The C# .Net code to respond to SSIS framework application package failure should appear similar to Figure 6-71.

```
public void Main()
{
    // System::PackageName, System::TaskName
    // User::FailApplicationOnPackageFailure, User::PackagePath

    string packageName = Dts.Variables["System::PackageName"].Value.ToString();
    string taskName = Dts.Variables["System::TaskName"].Value.ToString();
    string subComponent = packageName + "." + taskName;
    int informationCode = 1001;
    int errorCode = -999;
    bool fireAgain = true;

    bool failApplicationOnPackageFailure = Convert.ToBoolean(Dts.Variables["User::FailApplicationOnPackageFailure"].Value);
    string packagePath = Dts.Variables["User::PackagePath"].Value.ToString();
    string description = String.Empty;

    if(failApplicationOnPackageFailure)
    {
        description = packagePath + " execution failed and FailApplicationOnPackageFailure is set (true)";
        Dts.Events.FireError(errorCode, subComponent, description, "", 0);
    }
    else
    {
        description = packagePath + " execution failed and FailApplicationOnPackageFailure is NOT set (false)";
        Dts.Events.FireInformation(informationCode, subComponent, description, "", 0, ref fireAgain);
    }

    Dts.TaskResult = (int)ScriptResults.Success;
}
```

Figure 6-71. *C# .Net code to log application package execution failure*

Close the VstaProjects window and click the OK button on the SCR Log Package Execution Failure script task editor.

Let's test it! A test execution results (still) in a failure on the part of this Parent.dtsx SSIS package execution, as shown in Figure 6-72.

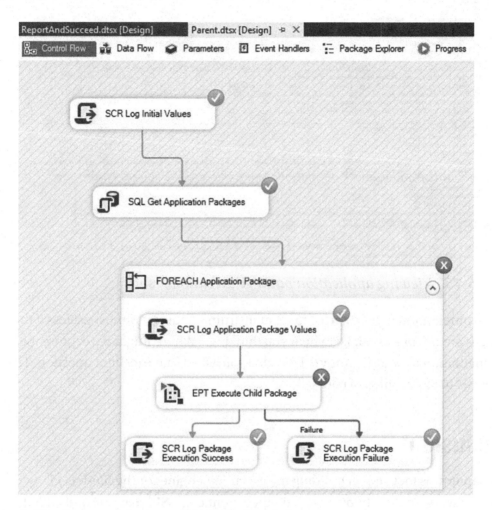

Figure 6-72. *Parent.dtsx execution fails (again)*

But this time, unlike previous executions, the information events instrumented in Parent.dtsx reveal what succeeded and what failed, as shown in Figure 6-73.

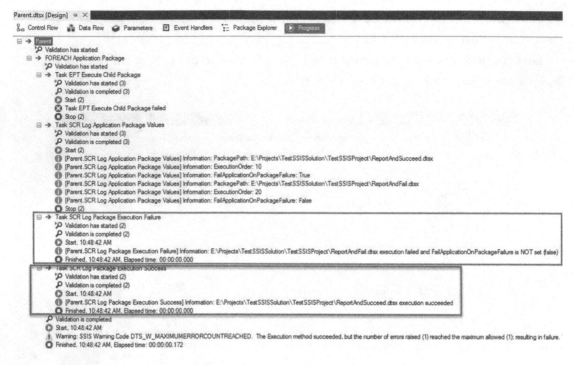

Figure 6-73. *Viewing application package execution results*

Instrumentation is good. Trust me. But instrumentation is almost useless if the messages are not persisted. I can hear you thinking, "Where might we capture these instrumentation messages, Andy?" I am glad you asked that excellent question. The answer is, "In SSISConfig, of course!"

Conclusion

This chapter was focused on building the execution engine for the file-based SSIS framework. In this chapter, we built the Parent.dtsx SSIS package and added instrumentation.

The next step is to record instrumentation messages in the SSISConfig database, which is covered in the next chapter.

CHAPTER 7

Framework Logging

In our journey to build a simple, custom, file-based SSIS framework, we have built a metadata database (named SSISDB), a test SSIS project, loaded some sample metadata (for the test SSIS project), created the execution engine (Parent.dtsx), and added instrumentation.

This chapter covers persisting instrumentation messages that are generated in the execution engine to the SSISConfig database.

Create a Log Schema

The information messages raised by the execution Parent.dtsx and displayed on the Progress/Execution Results tab are helpful, especially for troubleshooting. How may we store these messages for future access – especially when "something bad" has happened?

Return to your favorite T-SQL development editor (I am switching to Azure Data Studio version 1.17.1 for the remainder of this chapter), and connect to the SQL Server instance that hosts the SSISConfig database. Create a new schema named "log" in the SSISConfig database using the T-SQL in Listing 7-1.

Listing 7-1. Create the SSISConfig.log schema

```
Use [SSISConfig]
go

print 'Log schema'
If Not Exists(Select [schemas].[name]
              From [sys].[schemas]
                  Where [schemas].[name] = N'log')
  begin
```

© Andy Leonard, Kent Bradshaw 2020
A. Leonard and K. Bradshaw, *SQL Server Data Automation Through Frameworks*,
https://doi.org/10.1007/978-1-4842-6213-9_7

```
  print ' - Create log schema'
  declare @sql nvarchar(100) = N'Create Schema log'
  exec(@sql)
  print ' - Log schema created'
 end
Else
 begin
  print ' - Log schema already exists.'
 end
print ''
go
```

Messages appear a little different in Azure Data Studio, as shown in Figure 7-1.

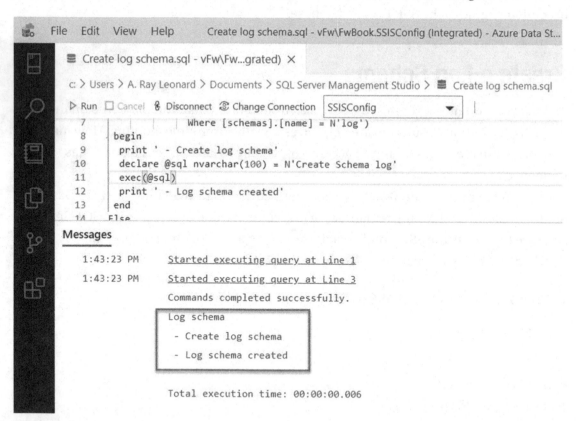

Figure 7-1. *Viewing messages from creating a log schema*

Messages we would expect in SQL Server Management Studio (SSMS) are highlighted in the "box" in Figure 7-1.

Before adding a table to capture "instances of application package execution" in the log schema, add a table to capture "instances of application execution" in the log schema. The first new table is named log.ApplicationInstance, and the T-SQL code to build this table is found in Listing 7-2.

Listing 7-2. Create the log.ApplicationInstance table in the SSISConfig database

```
use [SSISConfig]
go

print 'Log.ApplicationInstance table'
If Not Exists(Select [schemas].[name]
              + '.' + [tables].[name] As [Schema.Table]
              From [sys].[tables]
              Join [sys].[schemas]
                On [schemas].[schema_id] = [tables].[schema_id]
              Where [schemas].[name] = N'log'
                And [tables].[name] = N'ApplicationInstance')
 begin
  print ' - Create log.ApplicationInstance table'
  Create Table [log].[ApplicationInstance]
   (
     ApplicationInstanceId int identity(1, 1)
       Constraint PK_log_ApplicationInstance Primary Key Clustered
   , ApplicationId int Not NULL
       Constraint FK_log_ApplicationInstance_config_Applications
         Foreign Key References [config].[Applications](ApplicationId)
   , ApplicationStartTime datetimeoffset(7) Not NULL
       Constraint DF_log_ApplicationInstance_ApplicationStartTime
         Default(sysdatetimeoffset())
   , ApplicationEndTime datetimeoffset(7) NULL
   , ApplicationStatus nvarchar(25) Not NULL
      Constraint DF_log_ApplicationInstance_ApplicationStatus
        Default(N'Running')
   )
  print ' - Log.ApplicationInstance table created'
 end
```

```
Else
 begin
  print ' - Log.ApplicationInstance table already exists.'
 end
print ''
go
```

When the T-SQL in Listing 7-2 is executed in Azure Data Factory, the messages shown in Figure 7-2 are displayed.

Messages

2:43:02 PM	Started executing query at Line 1
2:43:02 PM	Started executing query at Line 3
	Commands completed successfully.
	Log.ApplicationInstance table
	- Create log.ApplicationInstance table
	- Log.ApplicationInstance table created

Figure 7-2. *Log.ApplicationInstance table creation messages*

A new record will be inserted into the log.ApplicationInstance table when an SSIS framework application starts executing in the Parent.dtsx package. When the SSIS framework application completes executing, the record will be updated. But wait, there's more!

The second new table is named log.ApplicationPackageInstance, and the T-SQL code to build this table is found in Listing 7-3.

Listing 7-3. Create the log.ApplicationPackageInstance table in the SSISConfig database

```
use [SSISConfig]
go

print 'Log.ApplicationPackageInstance table'
If Not Exists(Select [schemas].[name]
            + '.' + [tables].[name] As [Schema.Table]
            From [sys].[tables]
```

162

```
            Join [sys].[schemas]
              On [schemas].[schema_id] = [tables].[schema_id]
              Where [schemas].[name] = N'log'
                And [tables].[name] = N'ApplicationPackageInstance')
begin
 print ' - Create log.ApplicationPackageInstance table'
 Create Table [log].[ApplicationPackageInstance]
  (
     ApplicationPackageInstanceId int identity(1, 1)
        Constraint PK_log_ApplicationPackageInstance Primary Key Clustered
     , ApplicationInstanceId int Not NULL
        Constraint FK_log_ApplicationPackageInstance_log_
        ApplicationInstance
          Foreign Key References [log].[ApplicationInstance]
          (ApplicationInstanceId)
     , ApplicationPackageId int Not NULL
       Constraint FK_log_ApplicationPackageInstance_config_
       ApplicationPackages
        Foreign Key References [config].[ApplicationPackages]
        (ApplicationPackageId)
     , ApplicationPackageStartTime datetimeoffset(7) Not NULL
Constraint DF_log_ApplicationPackageInstance_ApplicationPackageStartTime
 Default(sysdatetimeoffset())
     , ApplicationPackageEndTime datetimeoffset(7) NULL
     , ApplicationPackageStatus nvarchar(25) Not NULL
       Constraint DF_log_ApplicationPackageInstance_
       ApplicationPackageStatus
          Default(N'Running')
  )
 print ' - Log.ApplicationPackageInstance table created'
end
```

```
Else
 begin
  print ' - Log.ApplicationPackageInstance table already exists.'
 end
print ''
go
```

When the T-SQL in Listing 7-3 is executed in Azure Data Factory, the messages shown in Figure 7-3 are displayed.

```
Messages

    2:56:44 PM      Started executing query at Line 1
    2:56:44 PM      Started executing query at Line 3
                    Commands completed successfully.
                    Log.ApplicationPackageInstance table
                      - Create log.ApplicationPackageInstance table
                      - Log.ApplicationPackageInstance table created
```

Figure 7-3. Log.ApplicationPackageInstance table creation messages

A new record will be inserted into the log.ApplicationPackageInstance table when an SSIS framework application package starts executing in the EPT Execute Application Package in the Parent.dtsx package's FOREACH Application Package foreach loop container. When the SSIS framework application completes execution, the record will be updated.

The "EPT" prefix stands for "Execute Package Task." Naming conventions in SSIS are important because the event model does not include a field that identifies "Executable Type." You can find an SSIS naming convention at ssis.tips/pages/naming.html.

Add Application Instance Logging to Parent.dtsx

Return to the Parent.dtsx SSIS package. Drag a new Execute SQL Task onto the control flow. Delete the precedence constraint between the SCR Log Initial Values script task and the SQL Get Application Packages execute SQL task. Connect new precedence constraints – the first between SCR Log initial Values script task and the new execute SQL task, and the second between the new Execute SQL Task and the SQL Get Application Packages execute SQL task. Rename the new Execute SQL Task "SQL Log Application Instance Start," as shown in Figure 7-4.

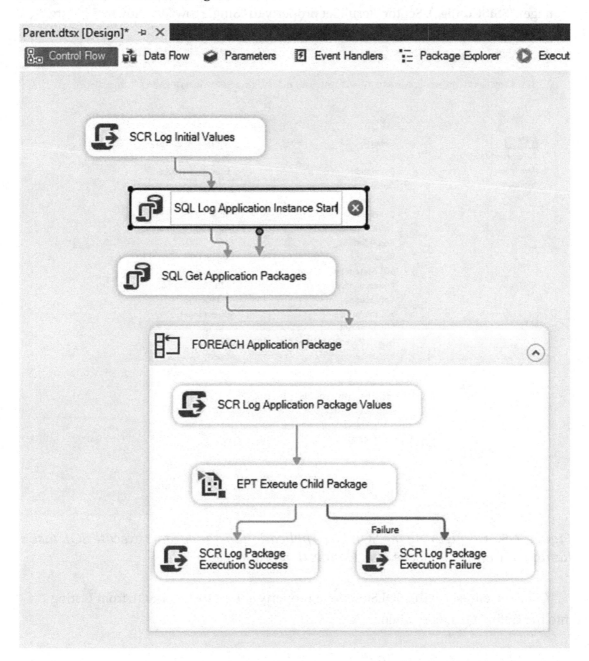

Figure 7-4. *Adding the SQL Log Application Instance Start execute SQL task*

Open the SQL Log Application Instance Start execute SQL task editor, and set the ConnectionType property to ADO.NET. Set the Connection property to the ADO.Net connection manager aimed at the SSISConfig database (I've renamed my connection manager "SSISConfig"). Set the ResultSet property to "Single row," as shown in Figure 7-5.

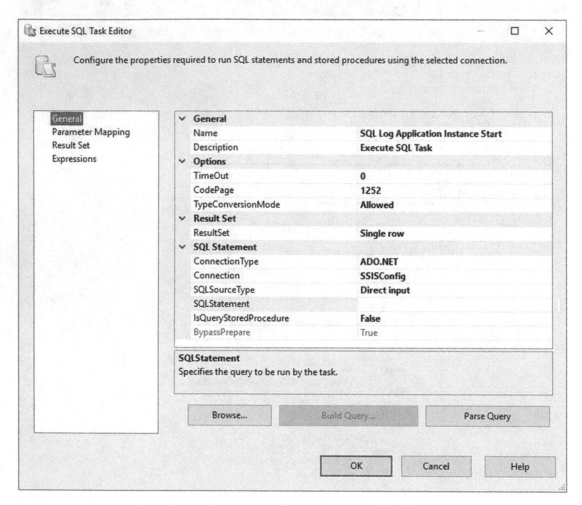

Figure 7-5. *Configuring the SQL Log Application Instance Start execute SQL task's ResultSet, ConnectionType, and Connection properties*

Click the ellipsis in the SQLStatement property and paste the T-SQL from Listing 7-4 into the Enter SQL Query dialog.

Listing 7-4. Inserting a row into the [log].[ApplicationInstance] table

```
declare @ApplicationId int = (Select ApplicationId
From config.Applications
Where ApplicationName = @ApplicationName)

Insert Into [log].ApplicationInstance (ApplicationId)
Output inserted.ApplicationInstanceId
Values (@ApplicationId)
```

The Enter SQL Query dialog should appear as shown in Figure 7-6.

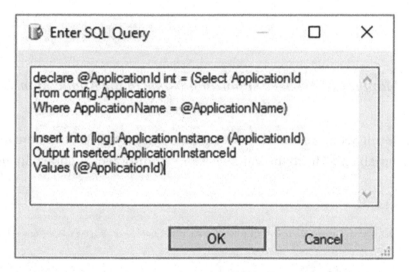

Figure 7-6. *Inserting a row into the [log].[ApplicationInstance] table*

Click the OK button to close the Enter SQL Query dialog.

Click the Parameter Mapping page and map the parameter
$Package::ApplicationName to the ApplicationName parameter in the SQLStatement
property as shown in Figure 7-7.

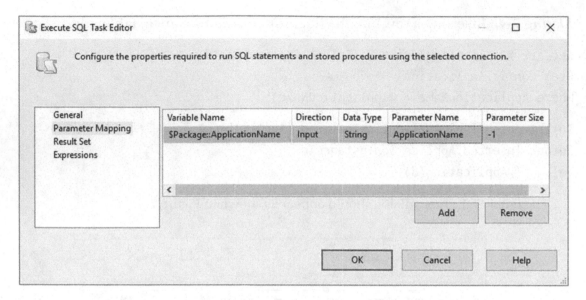

Figure 7-7. *Mapping $Package::ApplicationName to the ApplicationName query parameter*

Click the Result Set page, and click the Add button. Set the Result Name to "0" and select "<New variable...>" from the Variable Name drop-down, as shown in Figure 7-8.

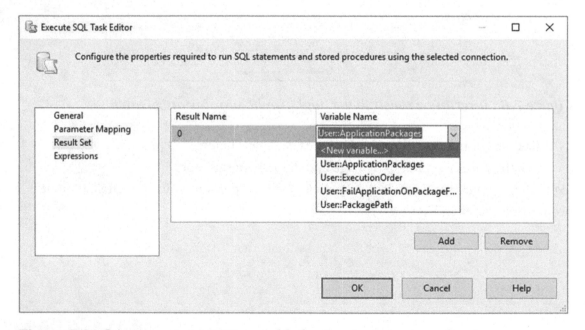

Figure 7-8. *Creating a new SSIS variable for the single-row result set*

When the Add Variable dialog displays, configure the new SSIS variable properties as shown in Figure 7-9:

- Container: Parent

- Name: ApplicationInstanceID

- Namespace: User

- Value type: Int32

- Value: -1

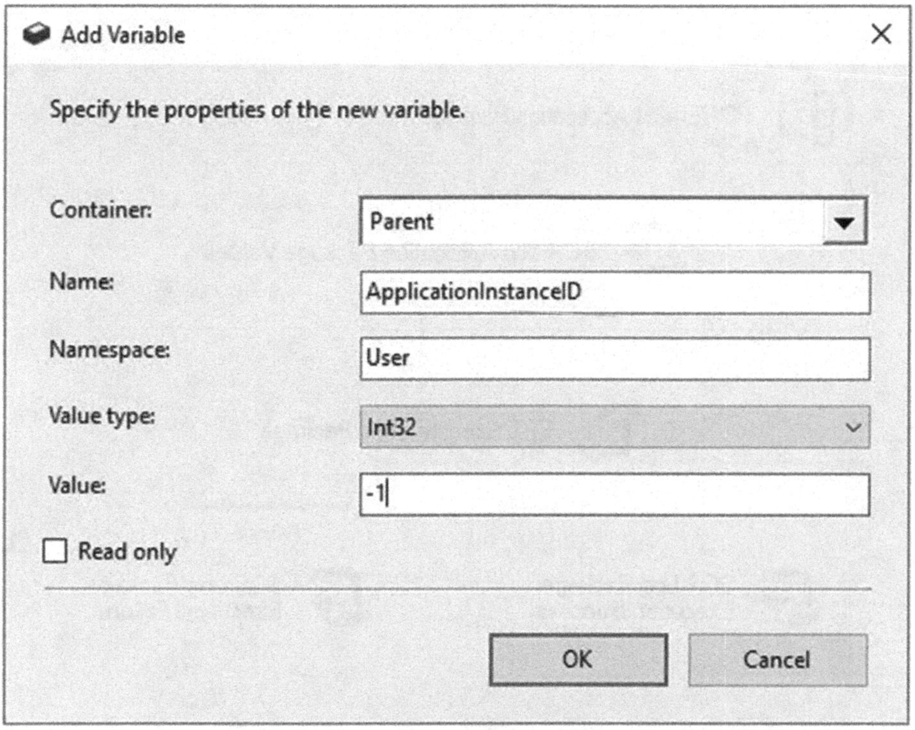

Figure 7-9. *Configuring the User::ApplicationInstanceID SSIS variable*

Click the OK button to close the Execute SQL Task Editor. The Parent.dtsx control flow should now appear as shown in Figure 7-10.

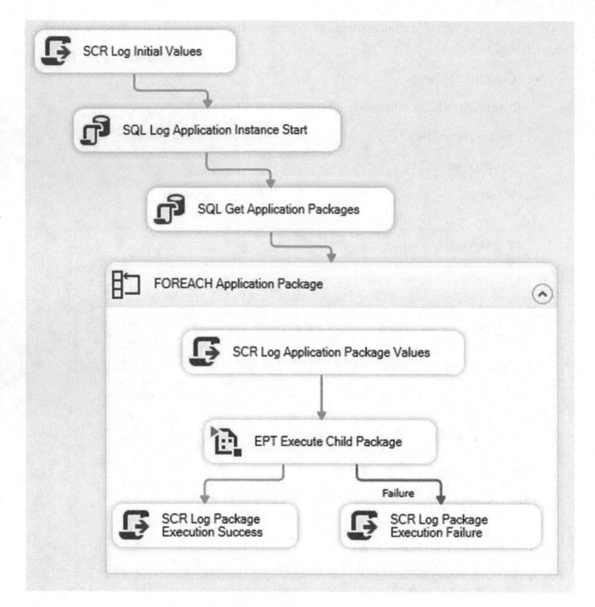

Figure 7-10. SQL Log Application Instance Start execute SQL task configured

Drag an Execute SQL Task onto the Parent.dtsx control flow below the FOREACH Application Package foreach loop container, and rename the Execute SQL Task to "SQL Log Application Instance Success," as shown in Figure 7-11.

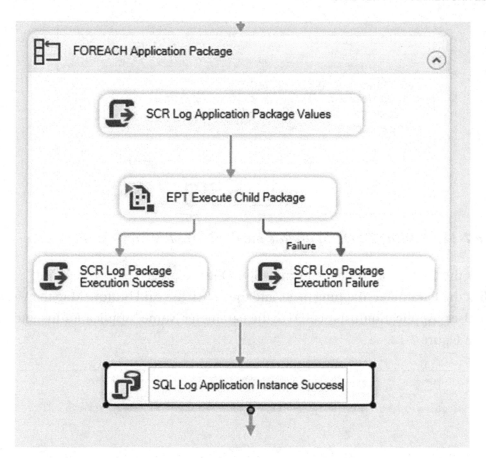

Figure 7-11. *Adding the SQL Log Application Instance Success execute SQL task*

Open the SQL Log Application Instance Success execute SQL task editor. Change the ConnectionType property to ADO.NET, and select the SSISConfig Connection. Click the ellipsis in the SQLStatement property. When the Enter SQL Query dialog displays, enter the T-SQL shown in Listing 7-5.

Listing 7-5. T-SQL to update the [log].[ApplicationInstance] table when the SSIS application succeeds

```
Update [log].ApplicationInstance
Set ApplicationEndTime = sysdatetimeoffset()
    , ApplicationStatus = 'Succeeded'
Where ApplicationInstanceId = @ApplicationInstanceId
```

The Enter SQL Query dialog should appear as shown in Figure 7-12.

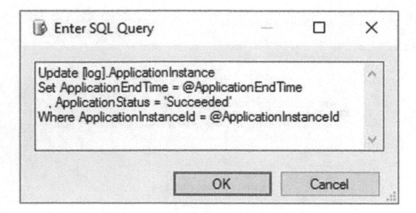

Figure 7-12. *Adding T-SQL to update the [log].[ApplicationInstance] table*

Click the OK button to close the Enter SQL Query dialog.

Click the Parameter Mapping page, and then click the Add button. Map the Variable Name "User::ApplicationInstanceID" to the Parameter Name "ApplicationInstanceId," as shown in Figure 7-13.

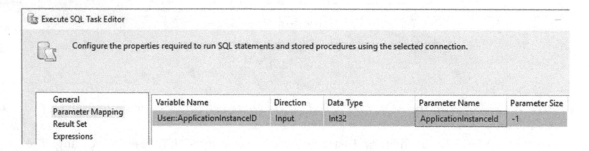

Figure 7-13. *Mapping the User:ApplicationInstanceID variable to the ApplicationInstanceId parameter*

Click the OK button to close the Execute SQL Task Editor.

The SQL Log Application Instance Success execute SQL task is now configured to update the row in the [log].[ApplicationInstance] table that was inserted when the SQL Log Application Instance Start execute SQL task executed earlier in the Parent.dtsx SSIS package, updating the row's ApplicationEndTime and ApplicationStatus fields to indicate the SSIS framework application instance completed successfully at such and such a time.

The next step is to update the ApplicationEndTime and ApplicationStatus fields in the row in [log].[ApplicationInstance] when the SSIS framework application instance *fails*.

Click the Event Handlers tab in the Parent.dtsx SSIS package, as shown in Figure 7-14.

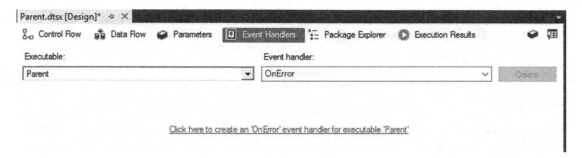

Figure 7-14. *The Parent.dtsx OnError Event handler*

If the Executable is not set to Parent, click the Executable drop-down, and select Parent as shown in Figure 7-15.

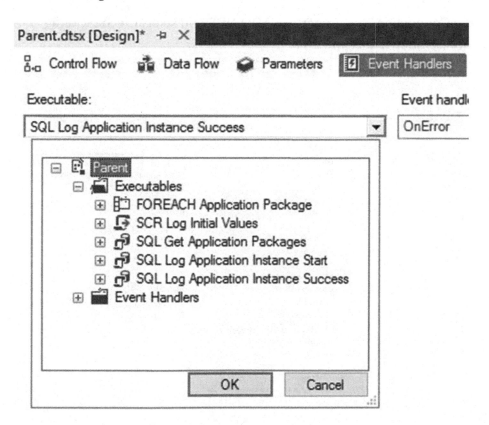

Figure 7-15. *Selecting the Parent Executable*

Click the link in the OnError Event handler labeled "Click here to create an 'OnError' event handler for the executable 'Parent'." Drag an Execute SQL Task onto the Event handler surface, and rename it "SQL Log Application Instance Failure," as shown in Figure 7-16.

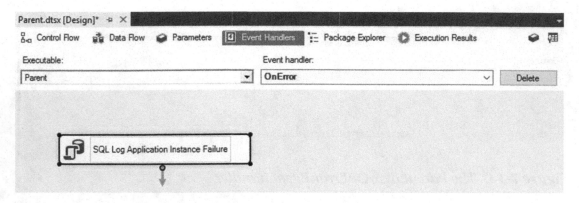

Figure 7-16. *Adding the SQL Log Application Instance Failure execute SQL task to the Parent OnError Event handler*

Set the ConnectionType property to ADO.NET, and select the SSISConfig Connection, as shown in Figure 7-17.

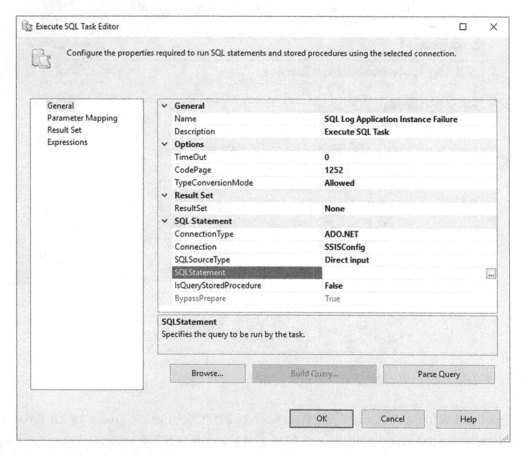

Figure 7-17. *Configuring the ConnectionType and Connection properties*

Click the ellipsis in the SQLStatement property to open the Enter SQL Query dialog, and enter the T-SQL shown in Listing 7-6 to update the [log].[ApplicationInstance] row when the SSIS application instance fails.

Listing 7-6. T-SQL to update [log].[ApplicationInstance] when the SSIS application fails

```
Update [log].ApplicationInstance
Set ApplicationEndTime = sysdatetimeoffset()
   , ApplicationStatus = 'Failed'
Where ApplicationInstanceId = @ApplicationInstanceId
```

The Enter SQL Query dialog should appear as shown in Figure 7-18.

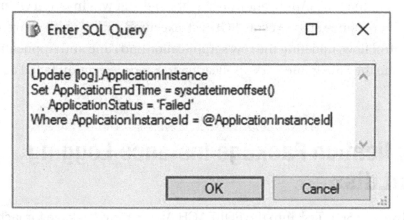

Figure 7-18. *Updating the [log].[ApplicationInstance] row*

Click the OK button to close the Enter SQL Query dialog.

Click the Parameter Mapping page, and then click the Add button. Map the Variable Name "User::ApplicationInstanceID" to the Parameter Name "ApplicationInstanceId," as shown in Figure 7-19.

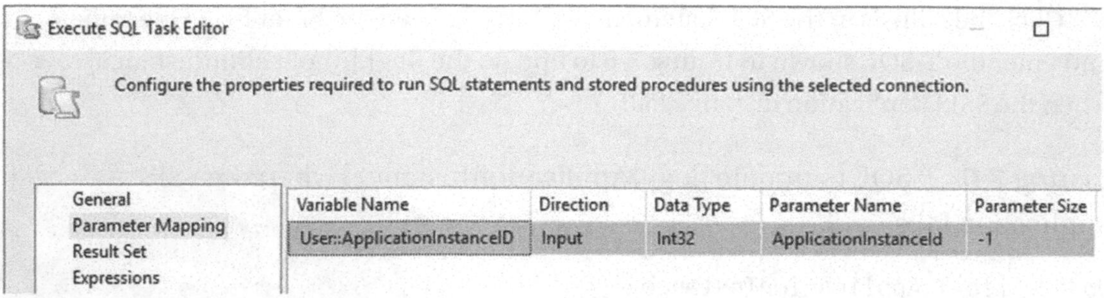

Figure 7-19. *Mapping the User:ApplicationInstanceID variable to the ApplicationInstanceId parameter*

Click the OK button to close the Execute SQL Task Editor.

The SQL Log Application Instance Failure execute SQL task is now configured to update the row in the [log].[ApplicationInstance] table that was inserted when the SQL Log Application Instance Start execute SQL task executed earlier on the Parent.dtsx SSIS package's control flow, updating the row's ApplicationEndTime and ApplicationStatus fields to indicate the SSIS framework application instance execution failed at such and such a time.

Add Application Package Instance Logging to Parent.dtsx

Drag a new Execute SQL Task into the FOREACH Application Package foreach loop container. Delete the precedence constraint between the SCR Log Application Package Values script task and the EPT Execute Child Package execute package task. Connect new precedence constraints – the first between SCR Log Application Package Values script task and the new execute SQL task, and the second between the new Execute SQL Task and the EPT Execute Child Package execute package task. Rename the new Execute SQL Task "SQL Log Application Package Instance Start," as shown in Figure 7-20.

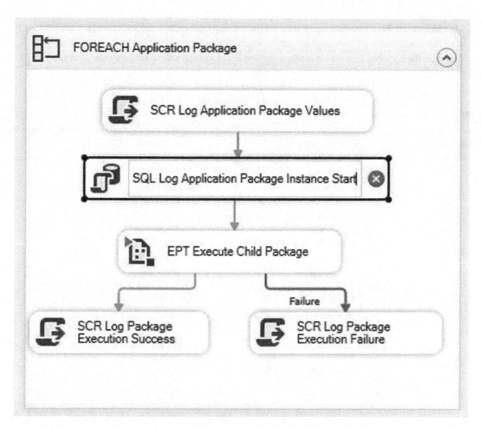

Figure 7-20. *Adding the SQL Log Application Package Instance Start execute SQL task*

Open the SQL Log Application Package Instance Start execute SQL task editor, and set the ConnectionType property to ADO.NET. Set the Connection property to the SSISConfig ADO.Net connection manager. Set the ResultSet property to "Single row," as shown in Figure 7-21.

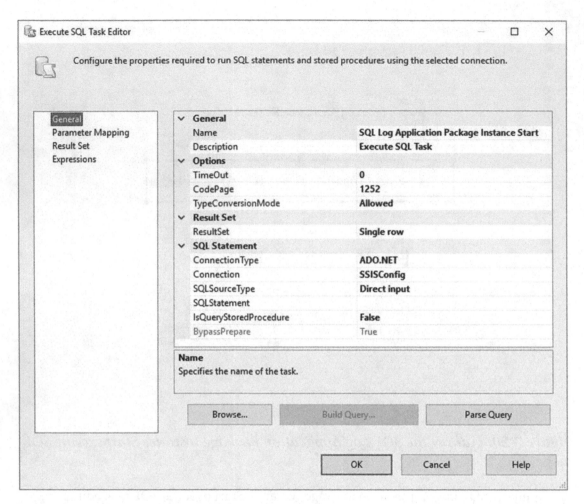

Figure 7-21. *Configuring the SQL Log Application Package Instance Start execute SQL task's ResultSet, ConnectionType, and Connection properties*

Click the ellipsis in the SQLStatement property, and paste the T-SQL from Listing 7-7 into the Enter SQL Query dialog.

Listing 7-7. Inserting a row into the [log].[ApplicationPackageInstance] table

```
declare @ApplicationPackageId int = (
  Select ap.ApplicationPackageId
  From config.ApplicationPackages ap
  Join config.Applications a
    On a.ApplicationId = ap.ApplicationId
  Join config.Packages p
```

```
  On p.PackageId = ap.PackageId
 Where ApplicationName = @ApplicationName
   And p.PackageLocation + p.PackageName = @PackagePath
)

Insert Into [log].ApplicationPackageInstance
(ApplicationInstanceId, ApplicationPackageId)
Output inserted.ApplicationPackageInstanceId
Values (@ApplicationInstanceId, @ApplicationPackageId)
```

The Enter SQL Query dialog should appear as shown in Figure 7-22.

Figure 7-22. *Inserting a row into the [log].[ApplicationPackageInstance] table*

Click the OK button to close the Enter SQL Query dialog.

Click the Parameter Mapping page, and map the following parameters (Variable Name, Data Type, Parameter Name):

- $Package::ApplicationName, String, ApplicationName

- User::PackagePath, String, PackagePath

- User::ApplicationInstanceID, Int32, ApplicationInstanceId

When complete, the Parameter Mapping page will appear as shown in Figure 7-23.

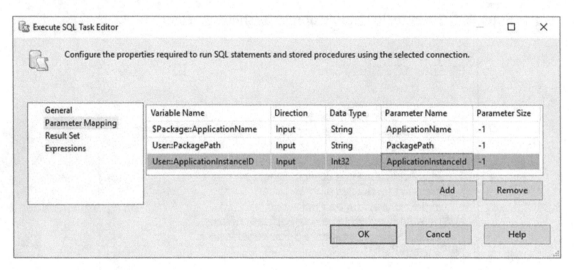

Figure 7-23. *Mapping ApplicationPackageInstance parameters*

Click the Result Set page, and click the Add button. Set the Result Name to "0" and select "<New variable...>" from the Variable Name drop-down, as shown in Figure 7-24.

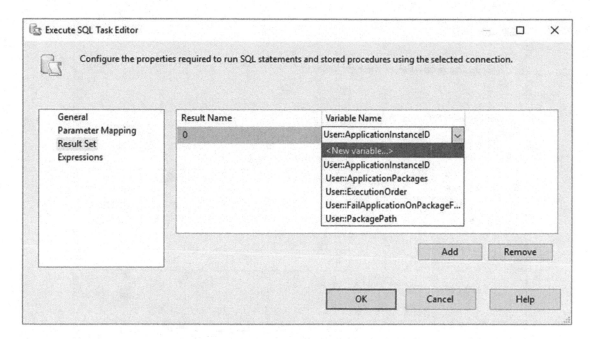

Figure 7-24. *Creating a new SSIS variable for the single-row result set*

When the Add Variable dialog displays, configure the new SSIS variable properties as shown in Figure 7-25.

- Container: Parent

- Name: ApplicationPackageInstanceID

- Namespace: User

- Value type: Int32

- Value: -1

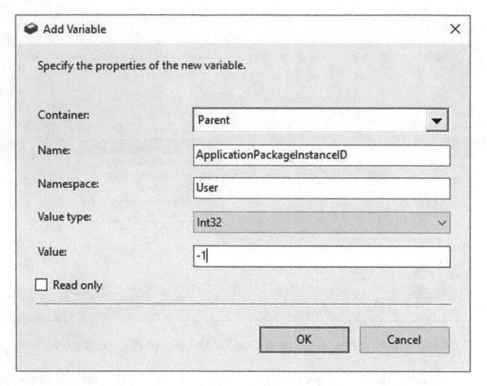

Figure 7-25. *Configuring the User::ApplicationPackageInstanceID SSIS variable*

Click the OK button to close the Execute SQL Task Editor. The FOREACH Application Package foreach loop container should now appear as shown in Figure 7-26.

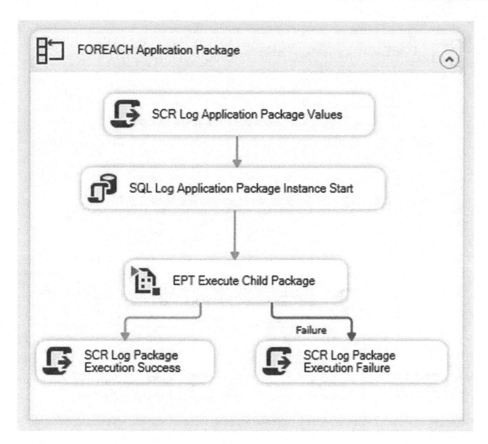

Figure 7-26. *SQL Log Application Package Instance Start execute SQL task configured*

Drag an Execute SQL Task onto the Parent.dtsx control flow below the SCR Log Package Execution Success script task in the FOREACH Application Package foreach loop container. Rename the Execute SQL Task "SQL Log Application Package Instance Success," as shown in Figure 7-27.

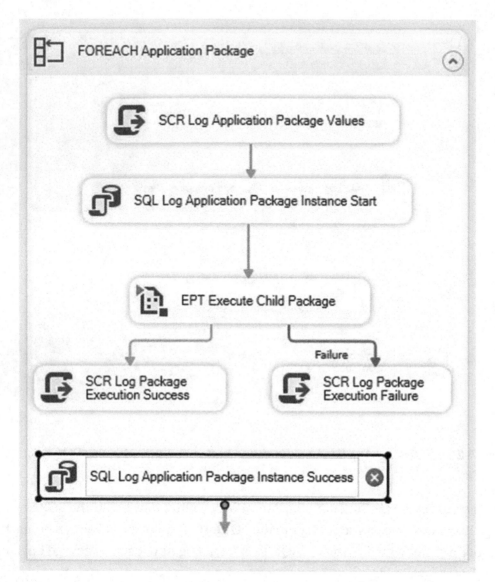

Figure 7-27. *Adding the SQL Log Application Package Instance Success execute SQL task*

Connect a precedence constraint from the SCR Log Package Execution Success script task to the SQL Log Application Package Instance Success execute SQL task. Open the SQL Log Application Package Instance Success execute SQL task editor. Change the ConnectionType property to ADO.NET, and select the SSISConfig Connection. Click the ellipsis in the SQLStatement property. When the Enter SQL Query dialog displays, enter the T-SQL shown in Listing 7-8.

Listing 7-8. T-SQL to update the [log].[ApplicationPackageInstance] table when the SSIS application package succeeds

```
Update [log].ApplicationPackageInstance
Set ApplicationPackageEndTime = sysdatetimeoffset()
   , ApplicationPackageStatus = 'Succeeded'
Where ApplicationPackageInstanceId = @ApplicationPackageInstanceId
```

The Enter SQL Query dialog should appear as shown in Figure 7-28.

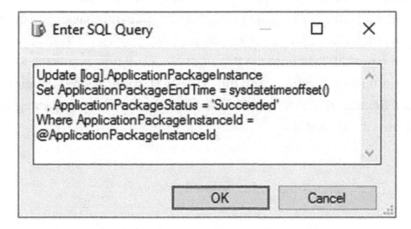

Figure 7-28. *Adding T-SQL to update the [log].[ApplicationPackageInstance] table*

Click the OK button to close the Enter SQL Query dialog.

Click the Parameter Mapping page, and then click the Add button. Map the Variable Name "User::ApplicationPackageInstanceID" to the Parameter Name "ApplicationPackageInstanceId," as shown in Figure 7-29.

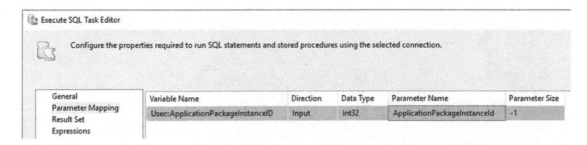

Figure 7-29. *Mapping the User:ApplicationPackageInstanceID variable to the ApplicationPackageInstanceId parameter*

Click the OK button to close the Execute SQL Task Editor.

The SQL Log Application Package Instance Success execute SQL task is now configured to update the row in the [log].[ApplicationPackageInstance] table that was inserted when the SQL Log Application Package Instance Start execute SQL task executed earlier in the FOREACH Application Package foreach loop container, updating the row's ApplicationPackageEndTime and ApplicationPackageStatus fields to indicate the SSIS framework application package instance completed successfully at such and such a time.

The next step is to update the ApplicationPackageEndTime and ApplicationPackageStatus fields in the row in [log].[ApplicationPackageInstance] when the SSIS framework application package instance *fails*.

Drag an Execute SQL Task onto the Parent.dtsx control flow below the SCR Log Package Execution Failure script task in the FOREACH Application Package foreach loop container. Rename the Execute SQL Task "SQL Log Application Package Instance Failure," as shown in Figure 7-30.

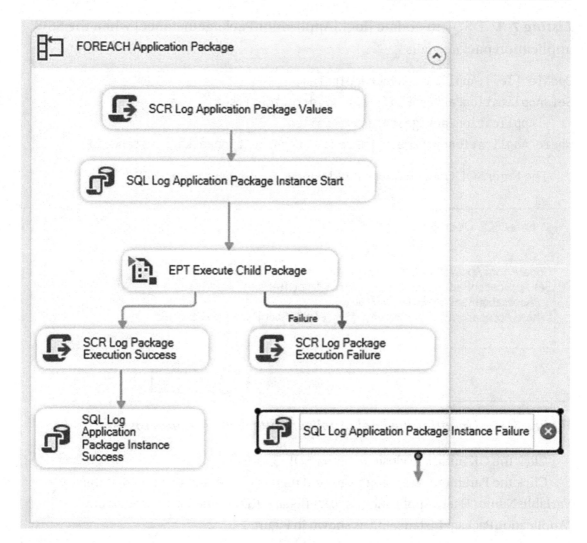

Figure 7-30. *Adding the SQL Log Application Package Instance Failure execute SQL task*

Connect a precedence constraint from the SCR Log Package Execution Failure script task to the SQL Log Application Package Instance Failure execute SQL task. Open the SQL Log Application Package Instance Failure execute SQL task editor. Change the ConnectionType property to ADO.NET, and select the SSISConfig Connection. Click the ellipsis in the SQLStatement property. When the Enter SQL Query dialog displays, enter the T-SQL shown in Listing 7-9.

Listing 7-9. T-SQL to update [log].[ApplicationPackageInstance] when the SSIS application package fails

```
Update [log].ApplicationPackageInstance
Set ApplicationPackageEndTime = sysdatetimeoffset()
  , ApplicationPackageStatus = 'Failed'
Where ApplicationPackageInstanceId = @ApplicationPackageInstanceId
```

The Enter SQL Query dialog should appear as shown in Figure 7-31.

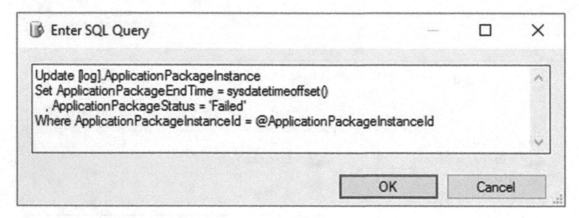

Figure 7-31. *Updating the [log].[ApplicationPackageInstance] row*

Click the OK button to close the Enter SQL Query dialog.

Click the Parameter Mapping page, and then click the Add button. Map the Variable Name "User::ApplicationPackageInstanceID" to the Parameter Name "ApplicationPackageInstanceId," as shown in Figure 7-32.

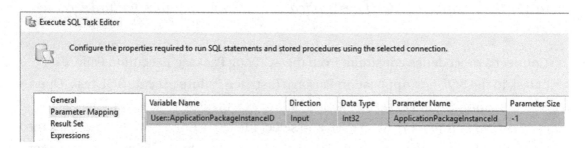

Figure 7-32. *Mapping the User:ApplicationPackageInstanceID variable to the ApplicationPackageInstanceId parameter*

Click the OK button to close the Execute SQL Task Editor.

The SQL Log Application Package Instance Failure execute SQL task is now configured to update the row in the [log].[ApplicationPackageInstance] table that was inserted when the SQL Log Application Package Instance Start execute SQL task executed earlier in the FOREACH Application Package foreach loop container, updating the row's ApplicationPackageEndTime and ApplicationPackageStatus fields to indicate the SSIS framework application package instance execution failed at such and such a time.

The Parent.dtsx SSIS package's FOREACH Application Package foreach loop container should appear similar to Figure 7-33.

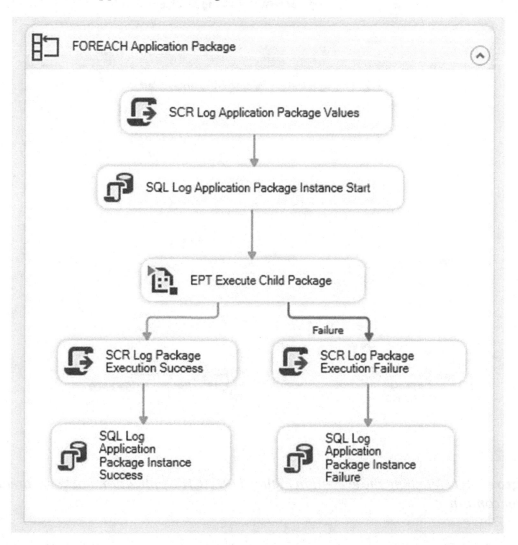

Figure 7-33. *The FOREACH Application Package foreach loop container configured to capture SSIS framework application package instrumentation*

Let's test it! Start the Parent.dtsx SSIS package in the debugger (press the F5 key). As currently configured, the SSIS framework application should execute and fail as shown in Figure 7-34.

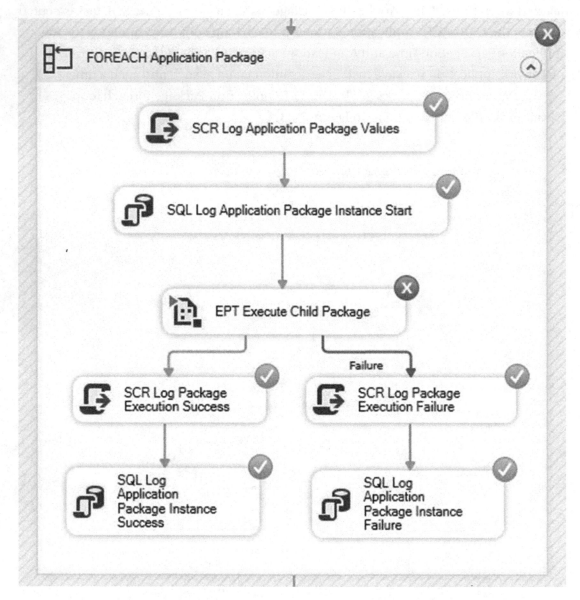

Figure 7-34. *Viewing the execution of the FOREACH Application Package foreach loop container*

The EPT Execute Child Package execute package task displayed in Figure 7-34 indicates failure. Two very important questions are

1. Did the execution test succeed?

2. Did the test execution succeed?

The answer to both questions lies in the log tables.

Viewing Execution Report Data

SSIS framework logging is stored in two tables name [log].[ApplicationInstance] and [log].[ApplicationPackageInstance]. Each row in the [log].[ApplicationInstance] table contains data about a single execution of an SSIS framework application or an *instance* of an application execution. View application instance data using the T-SQL query shown in Listing 7-10.

Listing 7-10. Viewing application instance log data

```
Select a.ApplicationName
    , ai.ApplicationStartTime
    , DateDiff(ms, ai.ApplicationStartTime, ai.ApplicationEndTime) As
ApplicationRunMilliSeconds
    , ai.ApplicationStatus
From [SSISConfig].[log].[ApplicationInstance] ai
Join [SSISConfig].[config].[Applications] a
  On a.ApplicationId = ai.ApplicationId
Order By ai.ApplicationStartTime Desc
```

Results from the application instance query will appear similar to those shown in Figure 7-35.

	ApplicationName	ApplicationStartTime	ApplicationRunMilliSeconds	ApplicationStatus
1	Framework Test	2020-05-22 08:50:06.9532116 ...	351	Failed
2	Framework Test	2020-05-21 09:42:47.1302552 ...	317	Failed
3	Framework Test	2020-05-19 16:32:07.6675319 ...	209	Failed

Results   Messages

Figure 7-35. *Viewing application instance results*

In answer to our first question – "Did the execution test succeed?" – the results of the SSIS application query indicate three test application instance executions are recorded, which means, "Yes, the execution tests succeeded three times."

Examining the application package instance data is next. Execute the T-SQL shown in Listing 7-11 to view application package instance data.

Listing 7-11. Viewing application package instance log data

```
Select  a.ApplicationName
     ,  p.PackageName
     ,  api.ApplicationPackageStartTime
     ,  ai.ApplicationStatus
     ,  api.ApplicationPackageStatus
     ,  api.ApplicationPackageStartTime
     ,  ap.FailApplicationOnPackageFailure
     ,  DateDiff(ms, api.ApplicationPackageStartTime, ➤ api.
ApplicationPackageEndTime) As ApplicationPackageRunMilliSeconds
     ,  ai.ApplicationStartTime
     ,  DateDiff(ms, ai.ApplicationStartTime, ai.ApplicationEndTime) As ➤
ApplicationRunMilliSeconds
From [SSISConfig].[log].[ApplicationPackageInstance] api
Join [SSISConfig].[log].[ApplicationInstance] ai
  On ai.ApplicationInstanceId = api.ApplicationInstanceId
Join [SSISConfig].[config].[ApplicationPackages] ap
  On ap.ApplicationPackageId = api.ApplicationPackageId
Join [SSISConfig].[config].[Applications] a
  On a.ApplicationId = ap.ApplicationId
Join [SSISConfig].[config].[Packages] p
  On p.PackageId = ap.PackageId
Order By api.ApplicationPackageStartTime Desc
```

Results from the application package instance query will appear similar to those shown in Figure 7-36.

	ApplicationName	PackageName	ApplicationPackageStartTime	ApplicationStatus	ApplicationPackageStatus	ApplicationPackageStartTime	FailApplicationOnPackageFailure	ApplicationPackageRunMi
1	Framework Test	ReportAndFail.dtsx	2020-05-22 08:50:07.1462…	Failed	Failed	2020-05-22 08:50:07.146275…	1	216
2	Framework Test	ReportAndSucceed.dtsx	2020-05-22 08:50:07.0141…	Failed	Succeeded	2020-05-22 08:50:07.014190…	1	102
3	Framework Test	ReportAndFail.dtsx	2020-05-21 09:42:47.3443…	Failed	Failed	2020-05-21 09:42:47.344349…	1	183
4	Framework Test	ReportAndSucceed.dtsx	2020-05-21 09:42:47.1052…	Failed	Succeeded	2020-05-21 09:42:47.195273…	1	121

Figure 7-36. *Viewing application package instance results*

In answer to our second question – "Did the test execution succeed?" – the results of the SSIS application package query indicate two test application instance executions have failed, and each test execution of the ReportAndFail.dtsx application package failed, which means, "Yes, the test executions succeeded two times." Wait a minute. Why are two failures considered a success? Two failures are considered a success because the ReportAndFail.dtsx application package is designed to fail and the FailApplicationOnPackageFailure bit is set to true (1).

Test the assertion by setting the FailApplicationOnPackageFailure bit for the ReportAndFail.dtsx application package to False (0) using the T-SQL in Listing 7-12.

Listing 7-12. Resetting the FailApplicationOnPackageFailure bit for ReportAndFail.dtsx

```
Select p.PackageName
    , ap.FailApplicationOnPackageFailure
From [SSISConfig].[config].[ApplicationPackages] ap
Join [SSISConfig].[config].[Packages] p
  On p.PackageId = ap.PackageId
Where p.PackageName = N'ReportAndFail.dtsx'

Update ap
Set ap.FailApplicationOnPackageFailure = 0
From [SSISConfig].[config].[ApplicationPackages] ap
Join [SSISConfig].[config].[Packages] p
  On p.PackageId = ap.PackageId
Where p.PackageName = N'ReportAndFail.dtsx'

Select p.PackageName
    , ap.FailApplicationOnPackageFailure
From [SSISConfig].[config].[ApplicationPackages] ap
Join [SSISConfig].[config].[Packages] p
  On p.PackageId = ap.PackageId
Where p.PackageName = N'ReportAndFail.dtsx'
```

Note the Select statements surface the values of the
FailApplicationOnPackageFailure bit values before and after the Update statement. Strive
to communicate intent in T-SQL statements; "Commands completed successfully" is *not*
enough feedback.

Results from the application package update will appear similar to those shown in
Figure 7-37.

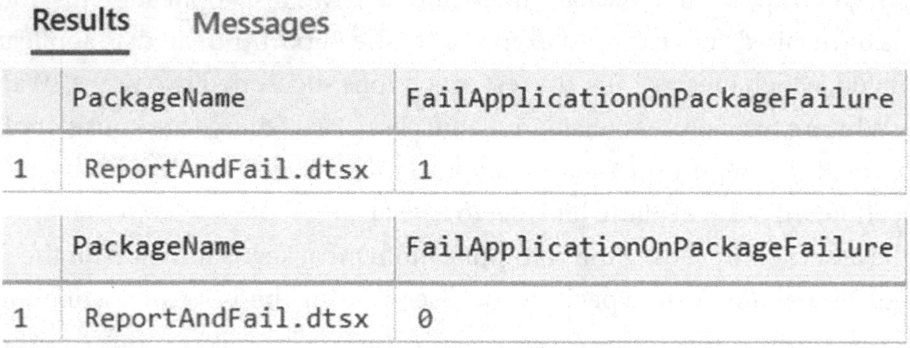

Results   Messages

	PackageName	FailApplicationOnPackageFailure
1	ReportAndFail.dtsx	1

	PackageName	FailApplicationOnPackageFailure
1	ReportAndFail.dtsx	0

Figure 7-37. *Viewing application package update results*

Executing Parent.dtsx in the SSIS debugger reveals *no change* in SSIS execution
results, as shown in Figure 7-38.

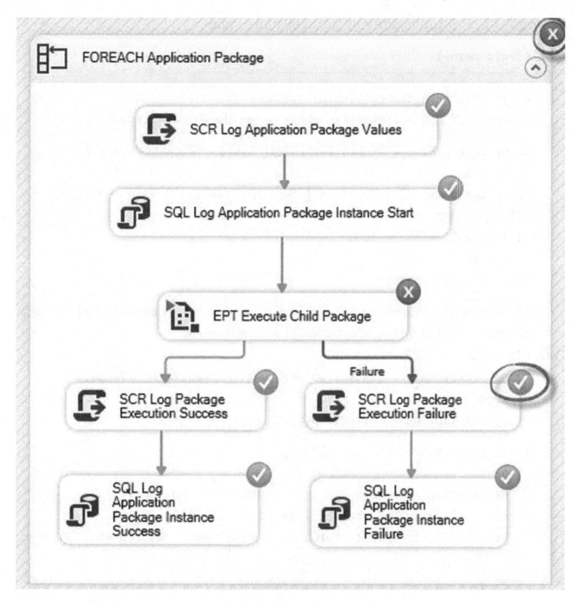

Figure 7-38. *Parent.dtsx execution fails, again s*

What gives? Fault tolerance for Parent.dtsx is not properly configured. Before we properly configure the fault tolerance for Parent.dtsx SSIS package, please note the FOREACH Application Package foreach loop container failed, indicating fault tolerance is not properly configured. Note also the SCR Log Package Execution Failure script task *succeeded*, indicating the FailApplicationOnPackageFailure logic – shown in Listing 7-13 – *is working*.

Listing 7-13. FailApplicationOnPackageFailure logic

```
public void Main()
  {
    // System::PackageName, System::TaskName
    // User::FailApplicationOnPackageFailure, User::PackagePath

    string packageName = ➤ Dts.Variables["System::PackageName"].Value.
    ToString();
    string taskName = Dts.Variables["System::TaskName"].Value.ToString();
    string subComponent = packageName + "." + taskName;
    int informationCode = 1001;
    int errorCode = -999;
    bool fireAgain = true;

    bool failApplicationOnPackageFailure = ➤ Convert.ToBoolean(Dts.Variabl
    es["User::FailApplicationOnPackageFailure"] ➤
.Value);
    string packagePath = Dts.Variables["User::PackagePath"].Value.
    ToString();
    string description = String.Empty;

    if(failApplicationOnPackageFailure)
      {
        description = packagePath + " execution failed and ➤
        FailApplicationOnPackageFailure is set (true)";
        Dts.Events.FireError(errorCode, subComponent, description, "", 0);
      }
      else
      {
        description = packagePath + " execution failed and ➤
        FailApplicationOnPackageFailure is NOT set (false)";
        Dts.Events.FireInformation(informationCode, subComponent, ➤
        description, "", 0, ref fireAgain);
      }

    Dts.TaskResult = (int)ScriptResults.Success;
  }
```

The Progress tab message from the SCR Log Package Execution Failure script task confirms FailApplicationOnPackageFailure functionality is working as designed, as shown in Figure 7-39.

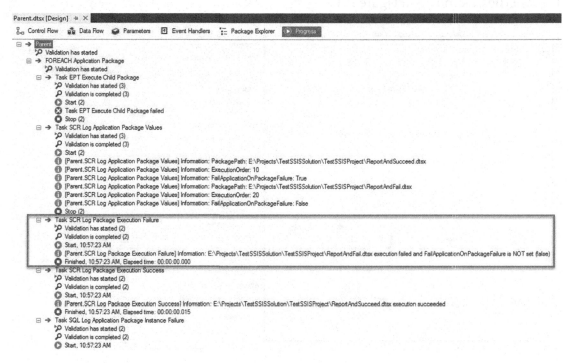

Figure 7-39. *FailApplicationOnPackageFailure is working*

The next step is to configure fault tolerance for the Parent.dtsx SSIS package.

Configure Fault Tolerance in SSIS Framework Application Metadata

Begin configuring fault tolerance for the Parent.dtsx SSIS package so that execution fails if and only if two conditions are met:

1. An application package execution fails.

2. The FailApplicationOnPackageFailure bit is set to true for the application package.

If one or both of these conditions are false, the desired behavior is that Parent. dtsx continues executing. To configure the *always-keep-running* functionality, click the FOREACH Application Package foreach loop container, and press the F4 key to display its properties, as shown in Figure 7-40.

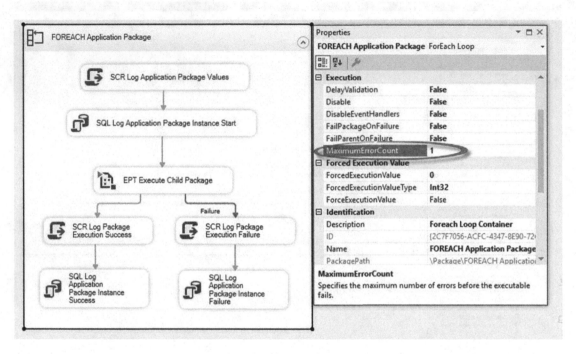

Figure 7-40. *Displaying properties of the FOREACH Application Package foreach loop container*

In Figure 7-40, the MaximumErrorCount property is highlighted. The net effect of the default MaximumErrorCount property value of 1 is as follows: if one error occurs within the FOREACH Application Package foreach loop container at execution time, the FOREACH Application Package foreach loop container execution will fail.

The FOREACH Application Package foreach loop container's MaximumErrorCount property may be increased but, if so, to what new value? Suppose 10, 20, or 300 application packages are configured to execute as part of an SSIS framework application. While there is actually a way to manage this very scenario, there exists a more elegant solution.

A more elegant solution is to ignore *all* the execution errors. Change the FOREACH Application Package foreach loop container's MaximumErrorCount property value to 0 to configure the FOREACH Application Package foreach loop container to ignore all errors during execution.

Credit is due to Julie Smith (@juliechix on Twitter) for teaching me about setting MaximumErrorCount to 0.

Because events *bubble* in SSIS, any error that occurs inside the FOREACH Application Package foreach loop container will "bubble up" to the Parent.dtsx SSIS package, which is the FOREACH Application Package foreach loop container's container. Click anywhere in the whitespace of the Parent.dtsx's control flow to display the Parent.dtsx package's properties, and change the MaximumErrorCount property to 0, as shown in Figure 7-41.

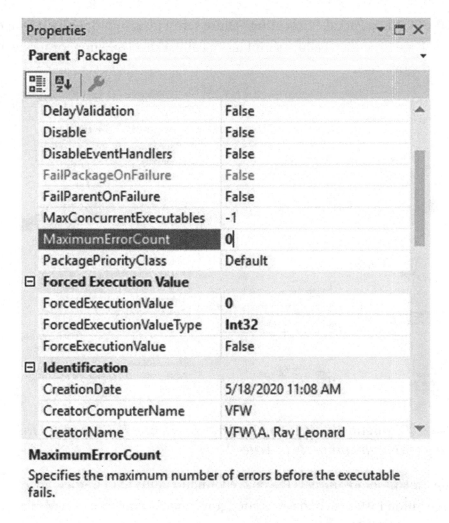

Figure 7-41. Changing the Parent.dtsx package's MaximumErrorCount property

As *currently* configured, SSIS framework application packages will continue to execute until the last application package is executed, regardless of whether the application package instance fails *and* regardless of the setting of the application package's FailApplicationOnPackageFailure bit.

The next step is to configure the Parent.dtsx SSIS package to *stop* executing if and only if two conditions are met:

1. An application package execution fails.

2. The FailApplicationOnPackageFailure bit is set to true for the application package.

Click the SCR Log Package Execution Failure script task. In properties, change the SCR Log Package Execution Failure script task's FailPackageOnFailure property to True, as shown in Figure 7-42.

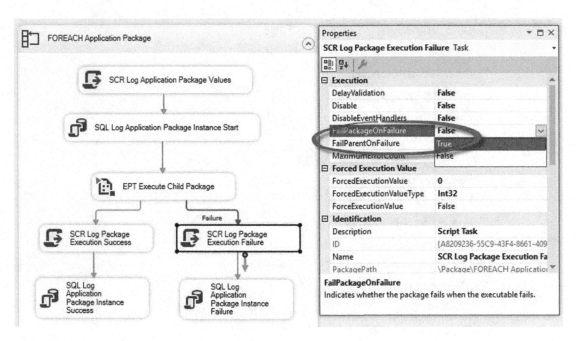

Figure 7-42. *Changing the SCR Log Package Execution Failure script task's FailPackageOnFailure property to True*

Testing execution now reveals the desired functionality; SSIS framework application package execution instances behave as configure, and fault tolerance is working as designed, as shown in Figure 7-43.

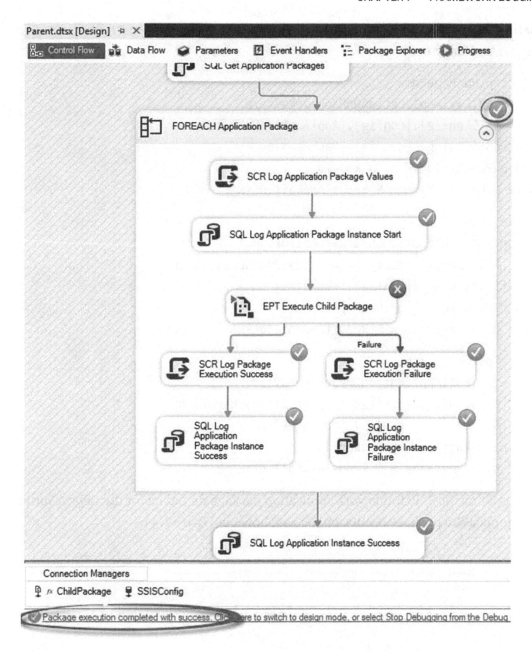

Figure 7-43. *SSIS application package fault tolerance working as designed*

Remember, the ReportAndFail.dtsx application package execution *should* fail, but the failure of a ReportAndFail.dtsx application package execution instance should *not* cause the application to fail (in its current configuration).

Update the FailApplicationOnPackageFailure bit configuration for the ReportAndFail.dtsx application package using the T-SQL in Listing 7-14.

Listing 7-14. Updating the FailApplicationOnPackageFailure bit for
ReportAndFail.dtsx

```
Select p.PackageName
     , ap.FailApplicationOnPackageFailure
From [SSISConfig].[config].[ApplicationPackages] ap
Join [SSISConfig].[config].[Packages] p
  On p.PackageId = ap.PackageId
Where p.PackageName = N'ReportAndFail.dtsx'

Update ap
Set ap.FailApplicationOnPackageFailure = 1
From [SSISConfig].[config].[ApplicationPackages] ap
Join [SSISConfig].[config].[Packages] p
  On p.PackageId = ap.PackageId
Where p.PackageName = N'ReportAndFail.dtsx'

Select p.PackageName
     , ap.FailApplicationOnPackageFailure
From [SSISConfig].[config].[ApplicationPackages] ap
Join [SSISConfig].[config].[Packages] p
  On p.PackageId = ap.PackageId
Where p.PackageName = N'ReportAndFail.dtsx'
```

Updating the SSIS framework application package metadata for the ReportAndFail.
dtsx application package results appears as shown in Figure 7-44.

Results   Messages

	PackageName	FailApplicationOnPackageFailure
1	ReportAndFail.dtsx	0

	PackageName	FailApplicationOnPackageFailure
1	ReportAndFail.dtsx	1

Figure 7-44. *Update ReportAndFail.dtsx application package metadata results*

Re-executing Parent.dtsx confirms we can fail if we want to, as shown in Figure 7-45.

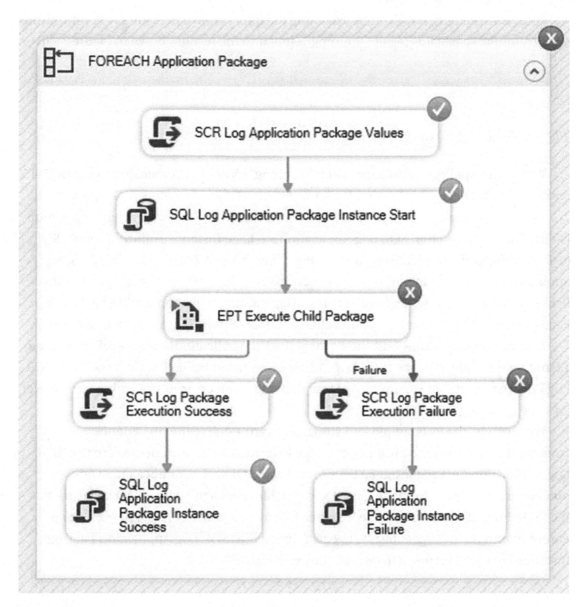

Figure 7-45. *Application Package Failure managed by SSIS framework metadata*

The current state of the Parent.dtsx SSIS package represents a fairly robust execution engine for the SSIS framework.

One issue remains, however: re-executing the query shown in Listing 7-11 returns the ApplicationPackageStatus result "Running," similar to the results shown in Figure 7-46.

ApplicationName	PackageName	ApplicationPackageStartTime	ApplicationStatus	ApplicationPackageStatus
Framework Test	ReportAndFail.dtsx	2020-05-22 11:55:16.8872218 …	Failed	Running

Figure 7-46. *ApplicationPaackageStatus is "Running"*

Why is the ApplicationPackageStatus "Running"? View the screenshot in Figure 7-45, and note the "SQL Log Application Package Instance Failure" execute SQL task *did not execute* because the precedence constraint connecting the "SCR Log Package Execution Failure" script task to the "SQL Log Application Package Instance Failure" execute SQL task is configured to evaluate on success *only*. Remember, looking at Listing 7-13, an application package failure may result in success *or* failure of the "SQL Log Application Package Instance Failure" execute SQL task, depending on the application package's FailApplicationOnPackageFailure bit value in the [config].[ApplicationPackages] table.

Listing 7-14 and Figure 7-44 show we set the FailApplicationOnPackageFailure bit to 1 (true), which informed the "SCR Log Package Execution Failure" script task. The "SCR Log Package Execution Failure" script task then raised an error, as prescribed by the .Net C# code in Listing 7-13.

Fortunately, the fix is simple and straightforward – delete the precedence constraint connecting the "SCR Log Package Execution Failure" script task to the "SQL Log Application Package Instance Failure" execute SQL task. Connect a new *failure* precedence constraint from the "EPT Execute Child Package" execute package task to the "SQL Log Application Package Instance Failure" execute SQL task. Re-executing Parent.dtsx in the debugger results in execution of the "SQL Log Application Package Instance Failure" execute SQL task, as shown in Figure 7-47.

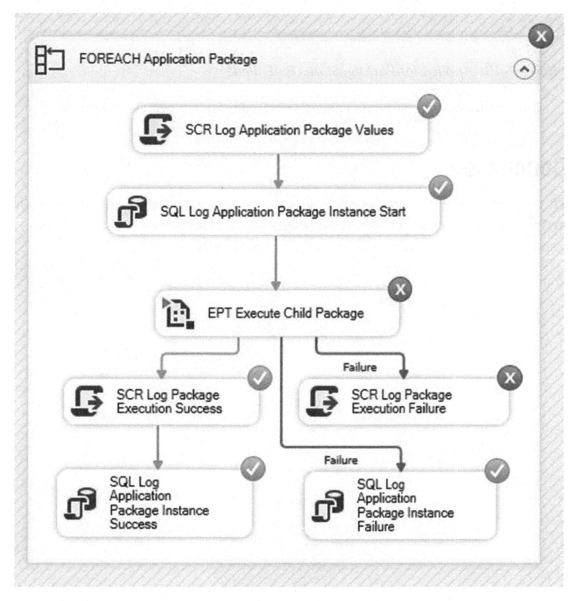

Figure 7-47. *The "SQL Log Application Package Instance Failure" execute SQL task executes*

Re-executing the application package status query in Listing 7-11 returns the ApplicationPackageStatus result "Failed," similar to the results shown in Figure 7-48.

ApplicationName	PackageName	ApplicationPackageStartTime	ApplicationStatus	ApplicationPackageStatus
Framework Test	ReportAndFail.dtsx	2020-06-01 13:05:16.7096071 …	Failed	Failed

Figure 7-48. *ApplicationPackageStatus is "Failed"*

Our SSIS framework execution engine is now ready for deployment elsewhere.

Conclusion

This chapter covered SSIS framework logging that extended the instrumentation we built in Chapter 6. The next few chapters cover migrating – and adapting – this simple SSIS framework to additional scenarios, including Azure Data Factory.

CHAPTER 8

Azure-SSIS Integration Runtime

The Microsoft Azure-SSIS team has been hard at work reducing the friction between executing SSIS packages on-premises and executing SSIS packages in Azure. Many enterprises execute SSIS packages stored in the file system on-premises. In the past, migrating SSIS packages executed in an on-premises file system meant changing enterprise SSIS storage and execution patterns. With the advent of Azure-SSIS File Share–based execution, executing SSIS packages from files in Azure is no longer an issue.

In fact, the SSIS framework from the previous chapter works well in Azure-SSIS. In this chapter, we will discuss and demonstrate

- Getting Started with Azure
- Provisioning an Azure Data Factory
- Provisioning Azure Storage
- Provisioning Azure-SSIS

Getting Started with Azure

Before creating and interacting with resources in Azure, one must first create an Azure account at azure.com, as shown in Figure 8-1.

© Andy Leonard, Kent Bradshaw 2020
A. Leonard and K. Bradshaw, *SQL Server Data Automation Through Frameworks*,
https://doi.org/10.1007/978-1-4842-6213-9_8

Figure 8-1. *Azure.com*

Please note the text on the link button in Figure 8-1 reads, "Try Azure for free." To date, the good people at Microsoft have consistently offered free introductory access to some Azure services. Free introductory offers lie solely with Microsoft and are subject to changes at Microsoft's discretion.

After an account has been created, the next step is to provision an Azure Data Factory.

Azure changes daily. Some of the screenshots and procedures in this book will be out of date before publication.

Provisioning an Azure Data Factory

Azure hosts lots of services. In general, Azure services may be categorized as

1. Infrastructure as a Service, or IaaS, in which Azure hosts enterprise infrastructure such as virtual machines (VM's).

2. Platform as a Service, or PaaS, in which Azure surfaces enterprise platforms such as SQL Server or Azure Data Factory.

Azure users begin by *provisioning* – or creating instances of – Azure service offerings. Begin provisioning an instance of Azure Data Factory – or ADF – by clicking "Create a resource" in the left menu, as shown in Figure 8-2.

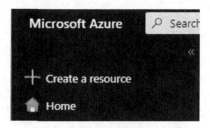

Figure 8-2. *Create a resource in Azure's left menu*

When the New page displays, search for "Data Factory," as shown in Figure 8-3.

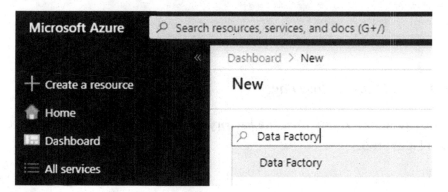

Figure 8-3. *Searching for Data Factory on the New page*

On the New page, click "Data Factory" beneath the search textbox on the New page to open the Data Factory page, as shown in Figure 8-4.

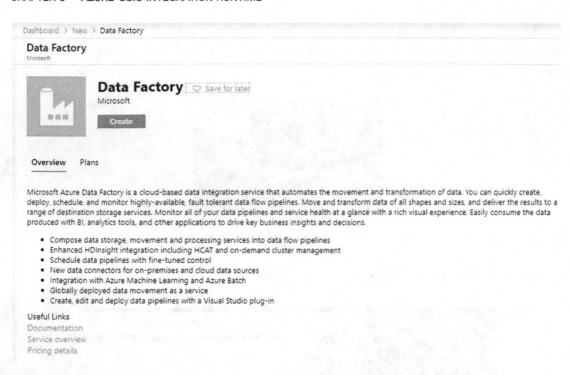

Figure 8-4. The Data Factory page

Click the Create button on the Data Factory page to open the New data factory page, as shown in Figure 8-5.

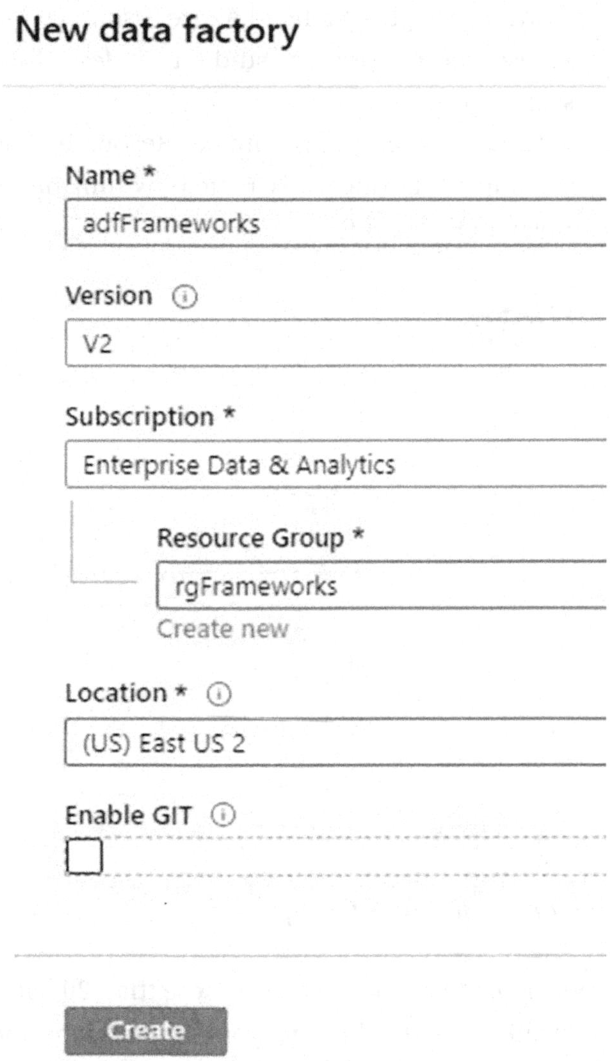

New data factory

Name *

adfFrameworks

Version ⓘ

V2

Subscription *

Enterprise Data & Analytics

Resource Group *

rgFrameworks

Create new

Location * ⓘ

(US) East US 2

Enable GIT ⓘ

☐

Create

Figure 8-5. *Adding a new instance of an Azure Data Factory*

Enter values for the required fields on the New data factory page.

Azure Data Factory, Resource Group, and Storage Account names are globally unique and case insensitive, according to Azure Data Factory – naming rules (docs. microsoft.com/en-us/azure/data-factory/naming-rules).

Resource groups are one way to group related Azure resources. When using Azure resources to learn Azure, resource groups allow students to *delete* the resource group and all constituents from a single screen.

You may select an existing resource group from the Resource Group drop-down, or create a new resource group by clicking the "Create new" link beneath the Resource Group drop-down, as shown in Figure 8-6.

New data factory

Name *
adftes

A resource group is a container that holds related resources for an Azure solution.

Version
V2

Name *
rgFrameworks

Subscri
Enterp

OK   Cancel

Create new

Location * ⓘ
(Europe) North Europe

Figure 8-6. *Creating a new Resource Group*

At the time of this writing, Microsoft maintains more than 20 data center locations around the globe. The locations exist to help get your applications and data closer to clients. Multiple locations also facilitate backups and redundancy.

For best response, select a location near you from the Locations drop-down, as shown in Figure 8-7.

Ne (Canada) Canada Central

(Europe) France Central

(Europe) North Europe

(Europe) UK South

(Europe) West Europe

(South America) Brazil South

(US) Central US

(US) East US

(US) East US 2

(US) North Central US

(Europe) North Europe

Figure 8-7. *Selecting a Location*

The "Enable GIT" is checked by default and this is awesome, and shown in Figure 8-8.

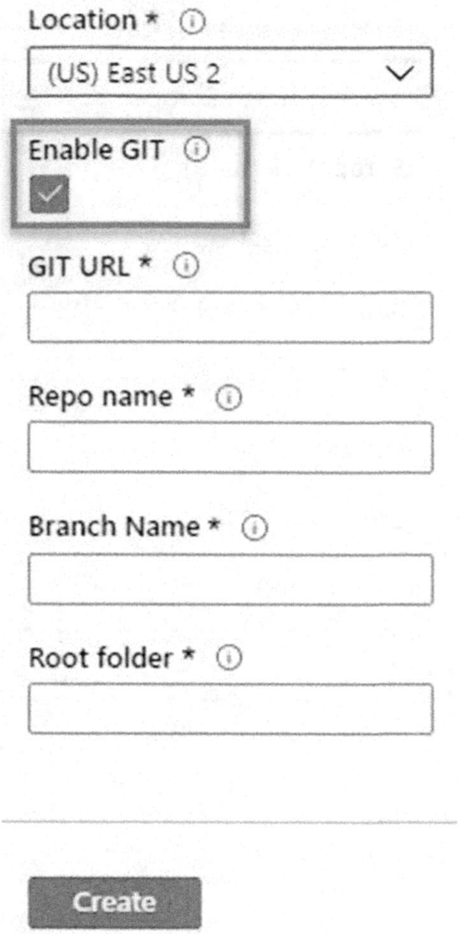

Location * ⓘ

(US) East US 2 ⌄

Enable GIT ⓘ
✓

GIT URL * ⓘ

Repo name * ⓘ

Branch Name * ⓘ

Root folder * ⓘ

Create

Figure 8-8. *The Enable GIT checkbox*

This chapter does not cover source control, so please uncheck the Enable GIT checkbox.

Click the Create button to provision an instance of ADF configured as specified.

After a few minutes, the Azure Data Factory is provisioned. You may visit the resource, or pin it to a dashboard using the buttons in the notification, as shown in Figure 8-9.

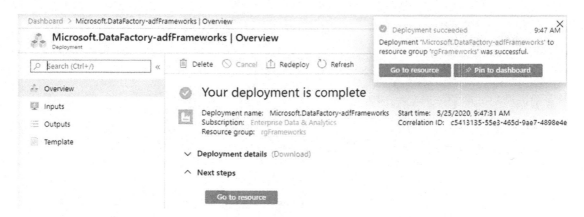

Figure 8-9. *Azure Data Factory, provisioned*

If you click the "Pin to dashboard" button from the notification "toast," a Data Factory tile appears on the last Azure dashboard you visited, and appears as shown in Figure 8-10.

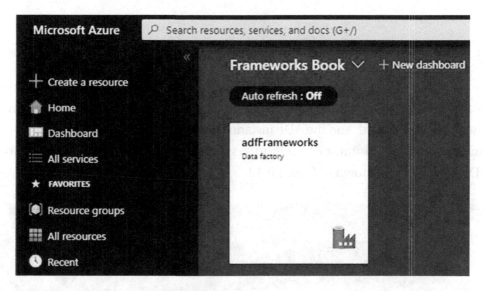

Figure 8-10. *An Azure Data Factory instance pinned to an Azure dashboard*

Like resource groups, Azure dashboards offer a way to group Azure resources. Azure dashboards allow visual grouping of Azure resources from many resource groups – and even subscriptions for those who manage more than one Azure subscription.

Click the tile to visit the page for the Azure Data Factory. The Overview displays by default, as shown in Figure 8-11.

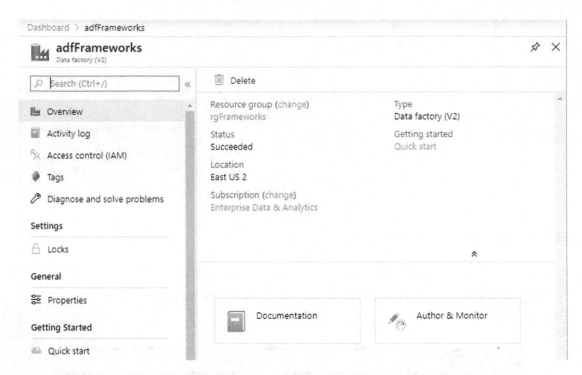

Figure 8-11. *The Azure Data Factory page*

The "pushpin" in the upper right corner of the Azure Data Factory page allows you to pin a tile for this ADF instance to an Azure Dashboard. Remember, you may configure more than one Azure dashboard, and the ADF instance tile will pin to the last dashboard visited.

Click the Author & Monitor link button to visit adf.azure.com for this instance of Azure Data Factory, as shown in Figure 8-12.

Figure 8-12. *The Data Factory Overview page*

At the time of this writing, the Data Factory Overview page includes shortcut links for the following operations:

- Create pipeline

- Create data flow

- Create pipeline from template

- Copy data

- Configure SSIS integration

- Set up code repository

Now that an Azure Data Factory is provisioned, the next step is to add Azure Storage.

Provisioning Azure Storage

Azure Blob Storage is a file system available to Azure resources and services. Like the file system on servers and laptops, Azure Blob Storage plays a vital role in all things Azure. The goal of this section is to describe provisioning a simple Azure Storage account. Later, we will use an Azure File Share in this account to store SSIS packages and ISPAC files. To learn more about Azure file share, please see docs.microsoft.com/en-us/azure/storage/files/storage-how-to-create-file-share.

As when provisioning an ADF instance, click Create a resource in the Azure left menu. Click Storage, and then click "Storage account – blob, file, table, queue," as shown in Figure 8-13.

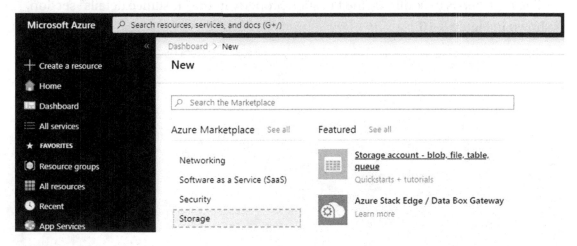

Figure 8-13. *Provisioning Azure Storage*

When the "Create storage account" page displays, configure the subscription and resource group in the "Project details" section, as shown in Figure 8-14.

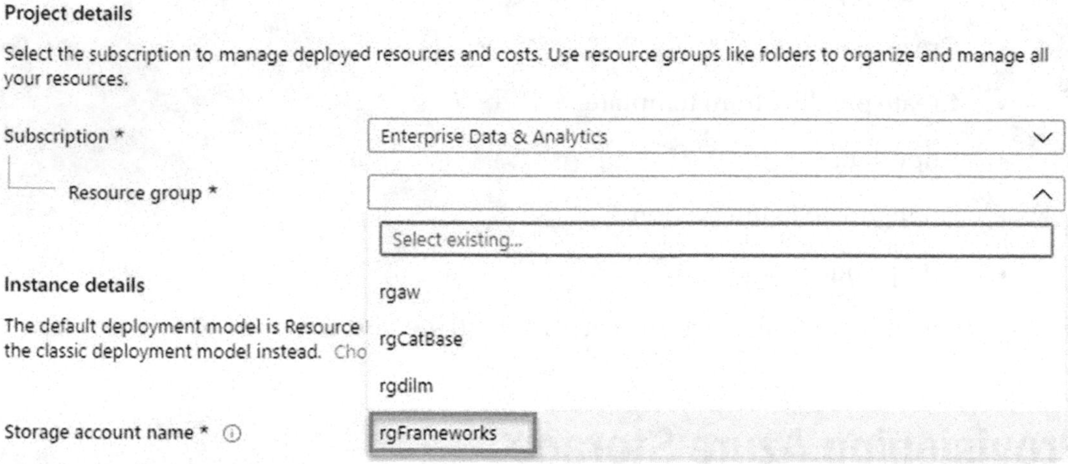

Project details

Select the subscription to manage deployed resources and costs. Use resource groups like folders to organize and manage all your resources.

Subscription *	Enterprise Data & Analytics	⌄

Resource group *		⌃

> Select existing...
>
> rgaw
>
> rgCatBase
>
> rgdilm
>
> rgFrameworks

Instance details

The default deployment model is Resource [ ] the classic deployment model instead. Cho

Storage account name * ⓘ

Figure 8-14. *Configuring the storage account subscription and resource group*

In the "Instance details" section, configure the "Storage account name" property, as shown in Figure 8-15.

Storage account name * ⓘ   stframeworks ✓

Figure 8-15. *Configuring the Storage account name property*

The next step is to configure the Location property in the "Instance details" section, as shown in Figure 8-16.

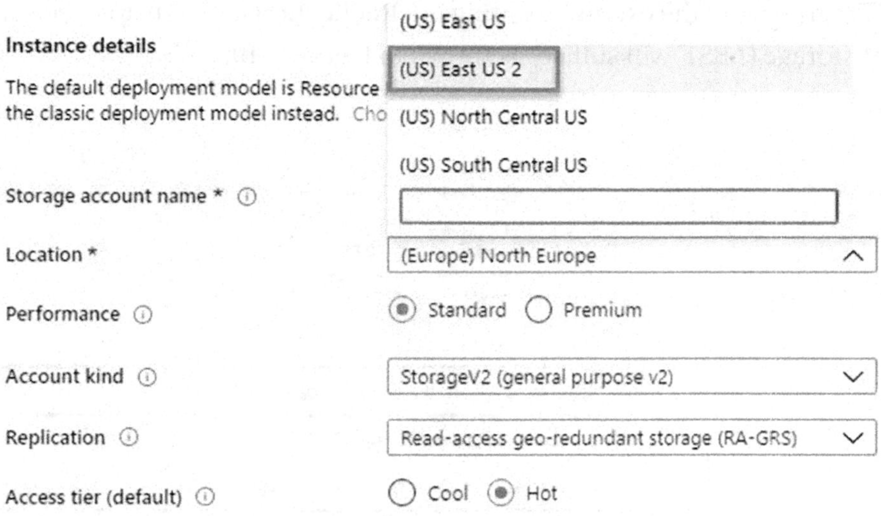

Instance details

The default deployment model is Resource
the classic deployment model instead. Cho

(US) East US

(US) East US 2

(US) North Central US

(US) South Central US

Storage account name * ⓘ

Location *   (Europe) North Europe   ∧

Performance ⓘ   ⦿ Standard ◯ Premium

Account kind ⓘ   StorageV2 (general purpose v2)   ∨

Replication ⓘ   Read-access geo-redundant storage (RA-GRS)   ∨

Access tier (default) ⓘ   ◯ Cool ⦿ Hot

Figure 8-16. *Configuring the Location property*

For the purposes of this example, "Standard" Performance will suffice, as shown in
Figure 8-17.

Performance ⓘ   ⦿ Standard ◯ Premium

Figure 8-17. *Configuring the Performance property*

Configure the "Account kind" property next. Set Account kind to "StorageV2 (general
purpose v2)," as shown in Figure 8-18.

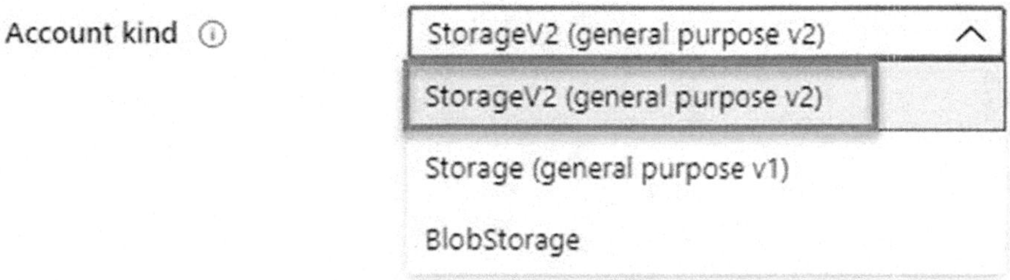

Account kind ⓘ   StorageV2 (general purpose v2)   ∧

StorageV2 (general purpose v2)

Storage (general purpose v1)

BlobStorage

Figure 8-18. *Configuring the Account kind property*

For the purposes of the example, setting the Replication property to "Locally-redundant storage (LRS)" will suffice, as shown in Figure 8-19.

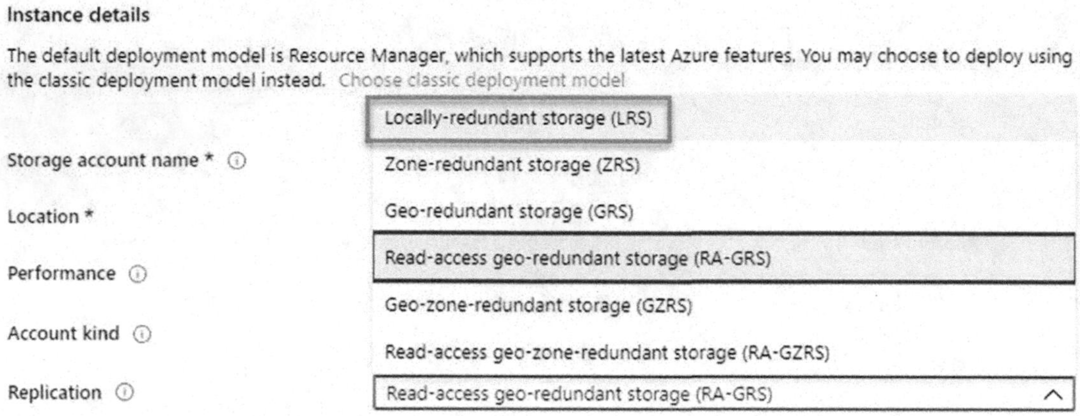

Instance details

The default deployment model is Resource Manager, which supports the latest Azure features. You may choose to deploy using the classic deployment model instead. Choose classic deployment model

	Locally-redundant storage (LRS)
Storage account name * ⓘ	Zone-redundant storage (ZRS)
Location *	Geo-redundant storage (GRS)
Performance ⓘ	Read-access geo-redundant storage (RA-GRS)
	Geo-zone-redundant storage (GZRS)
Account kind ⓘ	Read-access geo-zone-redundant storage (RA-GZRS)
Replication ⓘ	Read-access geo-redundant storage (RA-GRS) ∧

Figure 8-19. *Configuring the Replication property of the Storage account*

Accept the default "Access tier" property setting – "Hot" – as shown in Figure 8-20.

Access tier (default) ⓘ   ◯ Cool   ⦿ Hot

Figure 8-20. *Configuring the Access tier property*

The Basics tab of the "Create storage account" page is configured as shown in Figure 8-21.

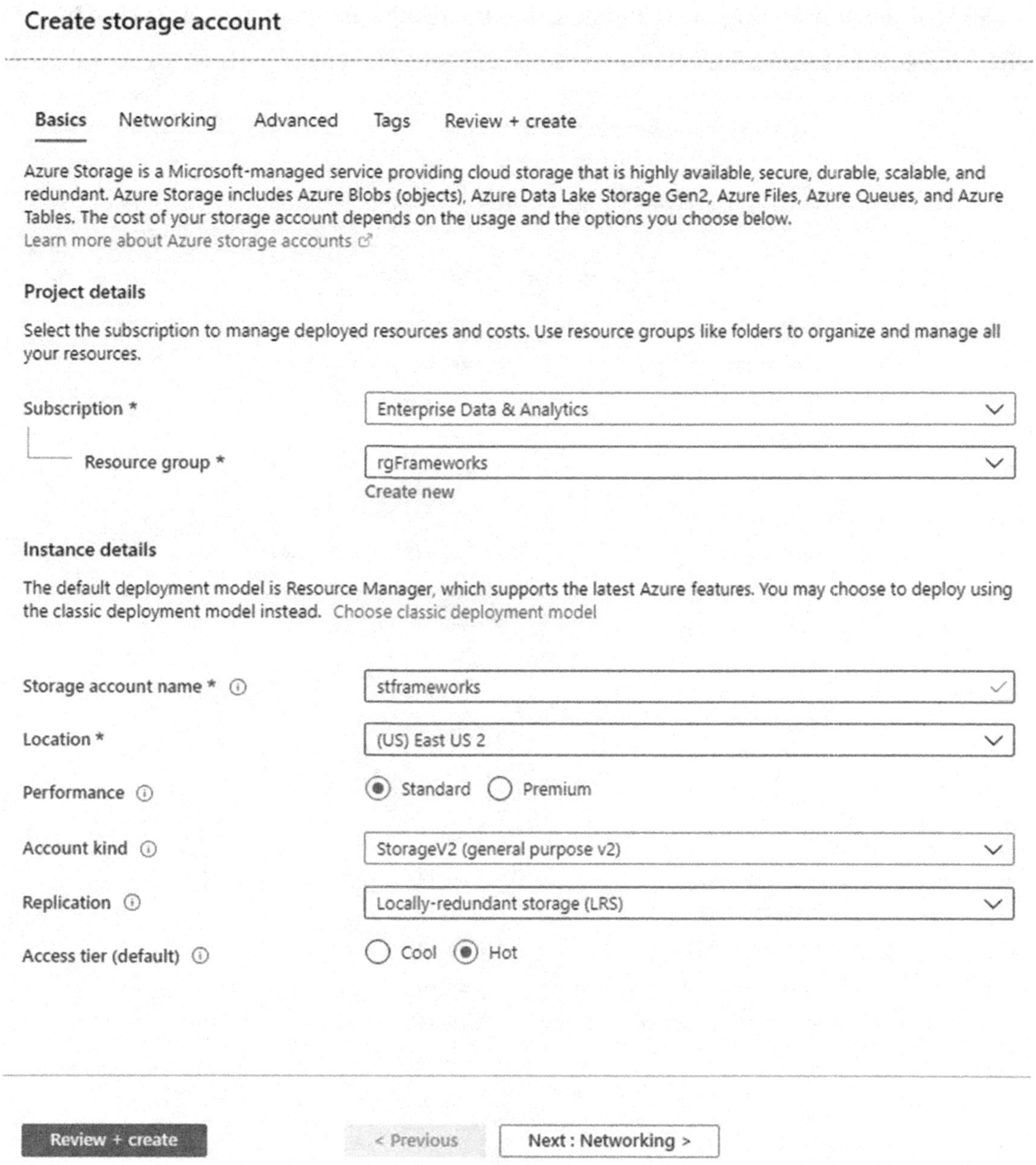

Create storage account

Basics   Networking   Advanced   Tags   Review + create

Azure Storage is a Microsoft-managed service providing cloud storage that is highly available, secure, durable, scalable, and redundant. Azure Storage includes Azure Blobs (objects), Azure Data Lake Storage Gen2, Azure Files, Azure Queues, and Azure Tables. The cost of your storage account depends on the usage and the options you choose below.
Learn more about Azure storage accounts

Project details

Select the subscription to manage deployed resources and costs. Use resource groups like folders to organize and manage all your resources.

Subscription *	Enterprise Data & Analytics ⌄
Resource group *	rgFrameworks ⌄
	Create new

Instance details

The default deployment model is Resource Manager, which supports the latest Azure features. You may choose to deploy using the classic deployment model instead. Choose classic deployment model

Storage account name * ⓘ	stframeworks ✓
Location *	(US) East US 2 ⌄
Performance ⓘ	⦿ Standard ◯ Premium
Account kind ⓘ	StorageV2 (general purpose v2) ⌄
Replication ⓘ	Locally-redundant storage (LRS) ⌄
Access tier (default) ⓘ	◯ Cool ⦿ Hot

Review + create   < Previous   Next : Networking >

Figure 8-21. *The Basics tab, configured*

For the purposes of this example, leave settings on the Networking, Advanced, and Tags tabs set to default. At the bottom of the Basics tab, click the "Review + create" button (shown in Figure 8-21).

Storage account settings are validated, and the results are displayed, as shown in Figure 8-22.

Figure 8-22. *Storage account configuration validation results*

Note also the text on the button at the bottom of the Create storage account page changes to "Create." Click the Create button to create the new Storage account. Once deployment is complete, the Storage account page and notification will display as shown in Figure 8-23.

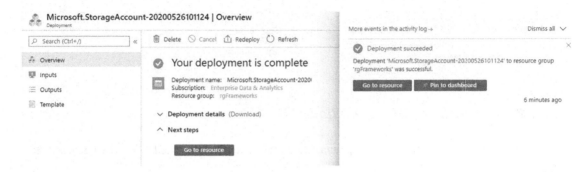

Figure 8-23. *Storage account, created*

Now that an Azure Data Factory and Storage account are provisioned, the next step is to add an Azure-SSIS integration runtime.

Provisioning an Azure-SSIS Integration Runtime for SSIS Package Files

On 30 June 2019, Microsoft released a new version of the ADF Azure-SSIS integration runtime. Enterprises could execute SSIS packages in the cloud previously using an SSIS Catalog hosted by an Azure database; the new release allowed SSIS package execution from an Azure file share.

To begin creating an Azure-SSIS integration runtime, navigate to an instance of Azure Data Factory's Manage page, and click "Integration runtimes" in the Connections category, as shown in Figure 8-24.

Data Factory
Author
Monitor
Manage
1
Data Factory
Connections
Linked services
Integration runtimes
2
Source control
Git configuration
Parameterization template

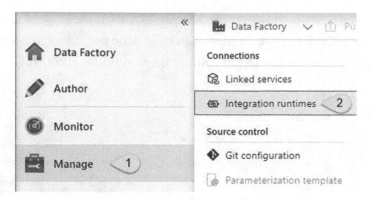

Figure 8-24. *Opening ADF Connections*

When the Connections page displays, click the "Integration runtimes" tab as shown in Figure 8-25.

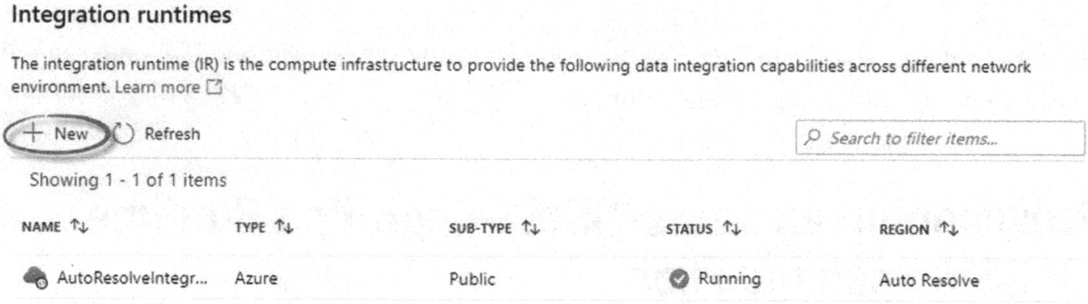

Integration runtimes
The integration runtime (IR) is the compute infrastructure to provide the following data integration capabilities across different network environment. Learn more
New
Refresh
Search to filter items...
Showing 1 - 1 of 1 items
NAME
TYPE
SUB-TYPE
STATUS
REGION
AutoResolveIntegr...
Azure
Public
Running
Auto Resolve

Figure 8-25. *ADF Connections "Integration runtimes" tab*

Click the "New" button – circled in Figure 8-25 – to open the "Integration runtime setup" blade and select the "Azure-SSIS" option as shown in Figure 8-26.

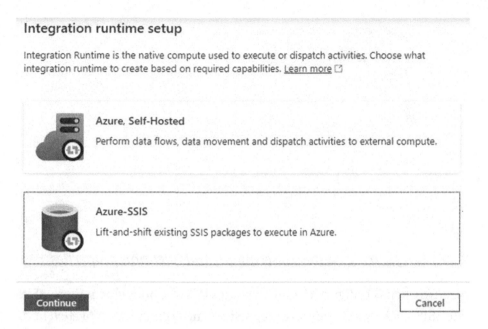

Figure 8-26. *Selecting Azure-SSIS in the Integration runtime setup blade*

The other option displayed in Figure 8-26 is "Azure, Self-Hosted." You may learn more about *all* Azure Data Factory integration runtimes by visiting docs.microsoft. com/en-us/azure/data-factory/concepts-integration-runtime#self-hosted-integration-runtime. For the purposes of this example, we will configure a "stand-alone" Azure-SSIS Integration Runtime – or IR – an IR that requires minimal security configuration. More information is available about joining an Azure-SSIS IR to a virtual network at docs. microsoft.com/en-us/azure/data-factory/join-azure-ssis-integration-runtime-virtual-network.

Click the Continue button to open the General settings page on the Integration runtime setup blade. Configure the name and optional description of the Azure-SSIS IR, and then set the Location property. For this example, set the Location property to the same location as the Azure Data Factory (which is the default), as shown in Figure 8-27.

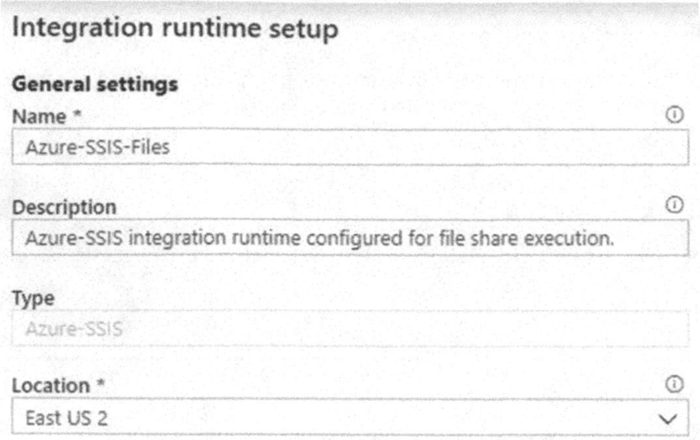

Integration runtime setup

General settings

Name *   ⓘ

Azure-SSIS-Files

Description   ⓘ

Azure-SSIS integration runtime configured for file share execution.

Type

Azure-SSIS

Location *   ⓘ

East US 2   ⌄

Figure 8-27. *Configuring Name, Description, and Location properties*

The next two steps configure virtual machines (VMs) upon which Azure-SSIS will execute SSIS packages. "Node size" represents the configuration of *each node* configured, and there are some beefy options, such as those listed in Figure 8-28.

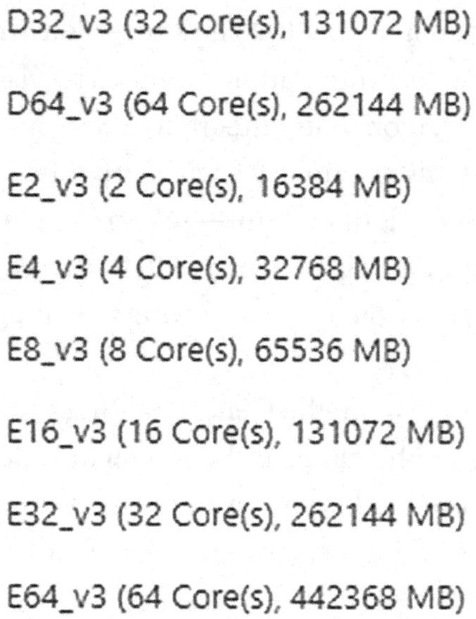

D32_v3 (32 Core(s), 131072 MB)

D64_v3 (64 Core(s), 262144 MB)

E2_v3 (2 Core(s), 16384 MB)

E4_v3 (4 Core(s), 32768 MB)

E8_v3 (8 Core(s), 65536 MB)

E16_v3 (16 Core(s), 131072 MB)

E32_v3 (32 Core(s), 262144 MB)

E64_v3 (64 Core(s), 442368 MB)

Figure 8-28. *Some Node size options for Azure-SSIS*

The first letter of the VM series indicates the general class and purpose of the virtual machine. A-Series VMs are specifically designed to save money during the testing and development phases of a project. To learn more about Azure virtual machine series and costs, please visit azure.microsoft.com/en-us/pricing/details/virtual-machines/series/.

226

Select the "A8_v2 (8 Core(s), 16384 MB)" Node size from the drop-down, as shown in Figure 8-29.

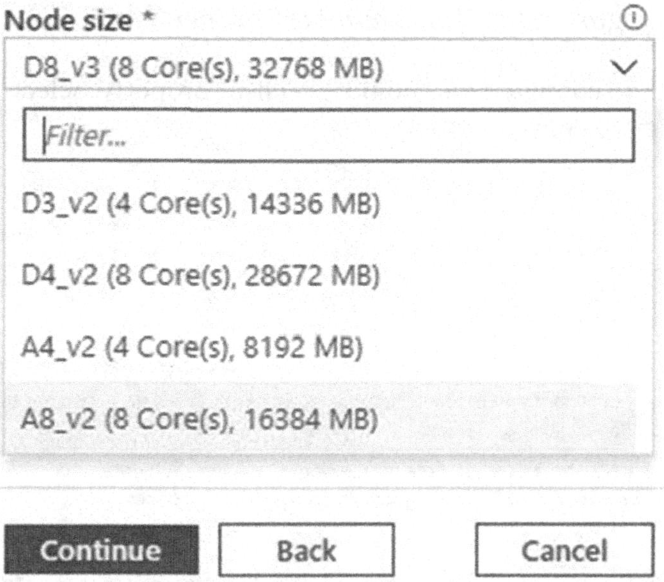

Figure 8-29. *Selecting a Node size for Azure-SSIS*

The "Node number" slider indicates the number of virtual machines available to Azure-SSIS. Each node will be configured according to the Node size selected in the previous drop-down. Node number is the number of nodes, configured in the Node size property, available to the Azure-SSIS IR. For this example, leave the Node number property set at 2 (the default), as shown in Figure 8-30.

Figure 8-30. *Configuring the "Node number" property to 2*

The next step is to configure the "Edition/license" property. Select "Standard" for now, as shown in Figure 8-31.

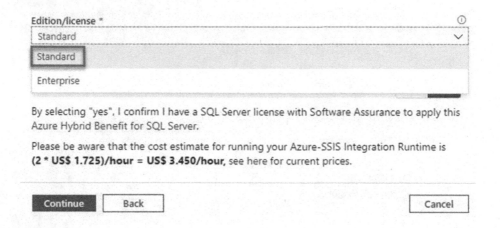

Figure 8-31. *Setting the Edition/license property*

A feature known as Bring Your Own License, or "BYOL," is another way to reduce spending n Azure. If you or your enterprise already own a SQL Server license, click the "Yes" to reduce the cost of running the VMs behind Azure-SSIS, as shown in Figure 8-32.

Node size *
A8_v2 (8 Core(s), 16384 MB)
Node number *
2
Edition/license *
Standard
Save money
Save with a license you already own. Already have a SQL Server license?
Yes
No
By selecting "yes", I confirm I have a SQL Server license with Software Assurance to apply this Azure Hybrid Benefit for SQL Server.
Please be aware that the cost estimate for running your Azure-SSIS Integration Runtime is (2 * US$ 0.945)/hour = US$ 1.890/hour, see here for current prices.

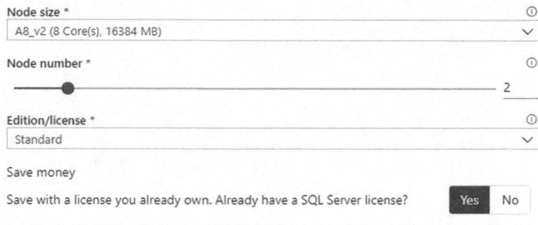

Figure 8-32. *Reduce Azure costs with BYOL*

Note changing the Edition/license property does *not* impact the Azure-SSIS cost, as shown in Figure 8-33.

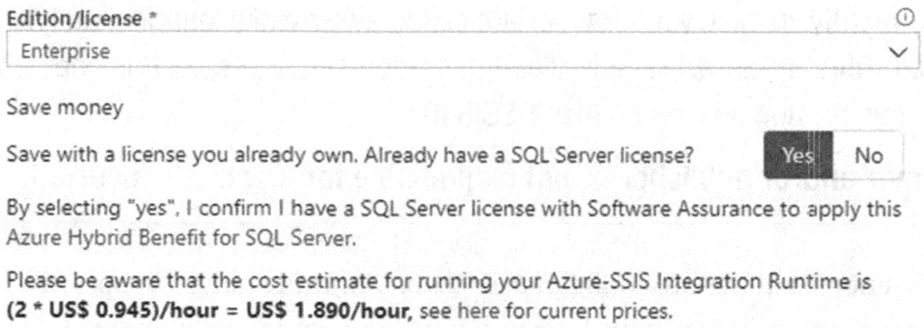

Edition/license *
Enterprise
Save money
Save with a license you already own. Already have a SQL Server license?
Yes
No
By selecting "yes", I confirm I have a SQL Server license with Software Assurance to apply this Azure Hybrid Benefit for SQL Server.
Please be aware that the cost estimate for running your Azure-SSIS Integration Runtime is (2 * US$ 0.945)/hour = US$ 1.890/hour, see here for current prices.

Figure 8-33. *Select Enterprise Edition does not impact Azure-SSIS cost*

Read azure.microsoft.com/en-us/pricing/details/data-factory/ssis/ to learn more about Azure-SSIS costs. You may also wish to try different variations to manage performance and costs, such as the "D2_v3 (2 Core(s), 8192 MB)" Enterprise configuration shown in Figure 8-34.

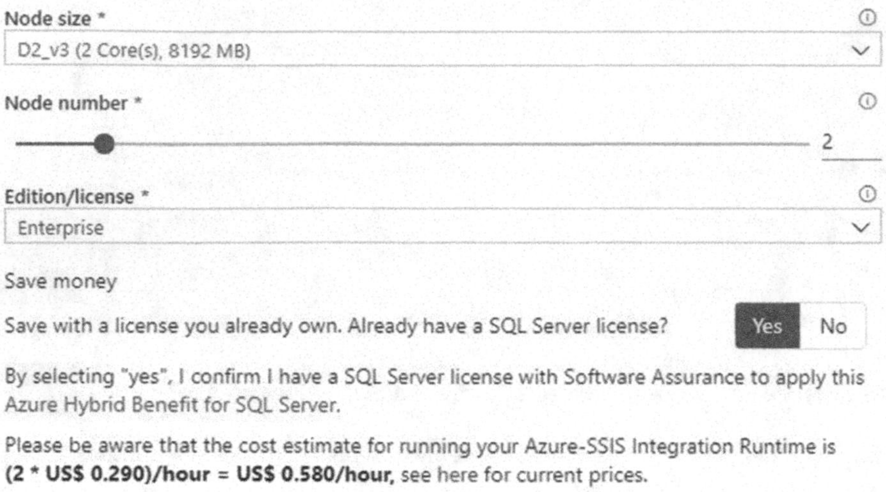

Node size *

D2_v3 (2 Core(s), 8192 MB)

Node number *

2

Edition/license *

Enterprise

Save money

Save with a license you already own. Already have a SQL Server license?   Yes   No

By selecting "yes", I confirm I have a SQL Server license with Software Assurance to apply this Azure Hybrid Benefit for SQL Server.

Please be aware that the cost estimate for running your Azure-SSIS Integration Runtime is **(2 * US$ 0.290)/hour = US$ 0.580/hour,** see here for current prices.

Figure 8-34. *Testing other combinations of Node size*

$0.58 is around 31 percent of $1.89.

Please visit azure.microsoft.com/en-us/pricing/details/data-factory/ssis/ to learn more about the cost of Azure-SSIS instances. One way to manage Azure-SSIS costs is to schedule an Azure Data Factory pipeline to shut down each Azure-SSIS instance nightly, in case you forget or are called away while learning. andyleonard. blog/2020/05/stop-an-azure-ssis-files-integration-runtime-safely/ is one detailed blog post about how to stop an Azure-SSIS IR.

The author and/or publisher is not responsible for any costs incurred.

Once General settings are configured, click the Continue button to open the "Deployment settings" page in the Integration runtime setup blade, as shown in Figure 8-35.

Integration runtime setup

Deployment settings

☑ Create SSIS catalog (SSISDB) hosted by Azure SQL Database server/Managed Instance to store ⓘ
your projects/packages/environments/execution logs
(See more info here)

Subscription *   ⓘ

| Enterprise Data & Analytics   ⌄ |

Location   ⓘ

| Select all   ⌄ |

Catalog database server endpoint *   ⓘ

| ⌄ |

☐ Use AAD authentication with the managed identity for your Data Factory ⓘ
(See how to enable it here)

Admin username *   ⓘ

| |

Admin username is empty

Admin password *   ⓘ

| 🔢 |

Admin password is empty

Catalog database service tier *   ⓘ

| S1   ⌄ |

☐ Create package stores to manage your packages that are deployed into file system/Azure   ⓘ
Files/SQL Server database (MSDB) hosted by Azure SQL Database Managed Instance
(See more info here)

───

| Test connection | | Back | | Cancel |

Figure 8-35. *The "Deployment settings" page in the Integration runtime setup*
blade

I can hear you thinking, "Gosh, Andy! That's a lot of stuff to configure. Plus, you wrote *nothing* about creating an Azure database. What gives?" Simmer down. We are creating an instance of Azure-SSIS for *files*, remember? Uncheck the "Create SSIS catalog (SSISDB) hosted by Azure SQL Database server/Managed Instance to store your projects/packages/environments/execution logs" checkbox – circled in Figure 8-35. The "Deployment settings" page in the Integration runtime setup blade now appears similar to that shown in Figure 8-36.

Integration runtime setup

Deployment settings

☐ Create SSIS catalog (SSISDB) hosted by Azure SQL Database server/Managed Instance to store   ⓘ
your projects/packages/environments/execution logs
(See more info here)

☐ Create package stores to manage your packages that are deployed into file system/Azure   ⓘ
Files/SQL Server database (MSDB) hosted by Azure SQL Database Managed Instance
(See more info here)

| Continue | Back | | Cancel |

Figure 8-36. *The "Deployment settings" page in the Integration runtime setup blade after unchecking the "Create SSIS catalog..." checkbox*

Click the Continue button to open the Integration runtime setup blade's Advanced settings page. Four options are available on the Advanced settings page:

1. Maximum parallel executions per node

2. Azure-SSIS integration runtime customization

3. VNet

4. Self-hosted integration runtime

The "Maximum parallel executions per node" drop-down does what it says. Each Azure-SSIS node (VM) may execute the number of SSIS packages configured by the Maximum parallel executions per node property.

Azure-SSIS integration runtimes may include custom configurations and/or third-party tools such as SentryOne's Task Factory (sentryone.com/products/task-factory) and SSIS+ Components Suite by COZYROC (cozyroc.com/products).

Most enterprises using the cloud store some, but not all, of their data in the cloud in a *hybrid* architecture. VNet is one way to access on-premises enterprise data. At the time of this writing, classic Azure virtual network is being deprecated and replaced with VNet. VNet is recommended for enterprises using or desiring to use

- Classic Azure virtual network

- Public IP addresses with Azure-SSIS IR

- Customized Azure-SSIS

Learn more by visiting the article "Join an Azure-SSIS integration runtime to a virtual network" at docs.microsoft.com/en-us/azure/data-factory/join-azure-ssis-integration-runtime-virtual-network.

A self-hosted Azure-SSIS integration runtime permits access to on-premises enterprise data *without* requiring VNet.

For the purposes of this example, accept the defaults – no checkboxes checked – as shown in Figure 8-37.

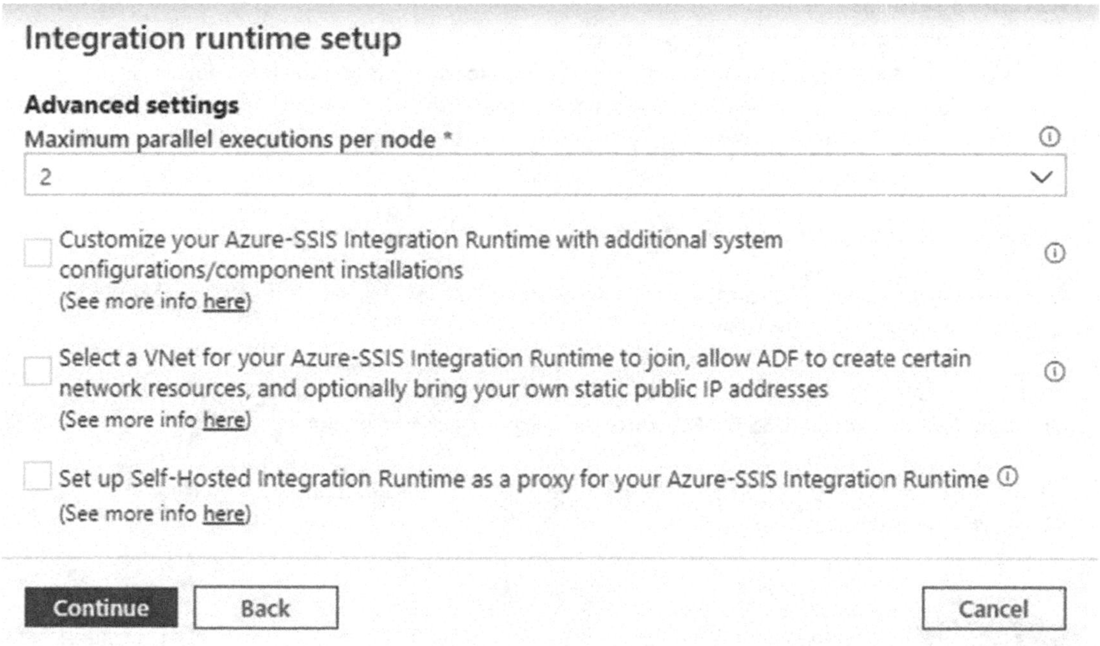

Figure 8-37. Configuring Azure-SSIS Advanced settings

Click the Continue button to proceed to the Integration runtime setup's Summary page. The Summary page displays Azure-SSIS configuration choices, as shown in Figure 8-38.

Integration runtime setup

Summary
Your Azure-SSIS Integration Runtime (IR) is created with the following settings:

Azure Data Factory Settings
- **Subscription:** 78ff08f6-334c-4e53-b737-8b5feaf74ecc
- **Resource group:** rgFrameworks
- **Name:** adfFrameworks
- **Location:** eastus2

General settings
- **Name:** Azure-SSIS-Files
- **Description:** Azure-SSIS integration runtime configured for file share execution.
- **Location:** East US 2
- **Node size:** Standard_D2_v3
- **Node number:** 2
- **Edition:** Enterprise
- **Azure Hybrid Benefit:** BasePrice

Advanced settings
- **Maximum parallel executions per node:** 2
- If you need to access data on premises, click **Previous** to do any of the followings:
  - Join your Azure-SSIS IR to a VNet connected to your on-premises network OR
  - Set up Self-Hosted Integration Runtime as a proxy for your Azure-SSIS Integration Runtime

If you want to change any of the above settings, click **Previous** to do so.

Once your Azure-SSIS IR is running, you can execute your packages on it after deploying them into your file system/Azure Files.

Please be aware that the cost estimate for running your Azure-SSIS Integration Runtime is **(2 * US$ 0.290)/hour = US$ 0.580/hour**, see here for current prices.

To manage the running cost of your Azure-SSIS IR, you can stop & restart it whenever convenient or schedule it just in time.

```
Create    Previous                                        Cancel
```

Figure 8-38. *Azure-SSIS configuration choices*

Click the Create button to provision the Azure-SSIS integration runtime. Connections ➤ Integration runtimes displays the new Azure-SSIS integration runtime in the "Starting" status, as shown in Figure 8-39.

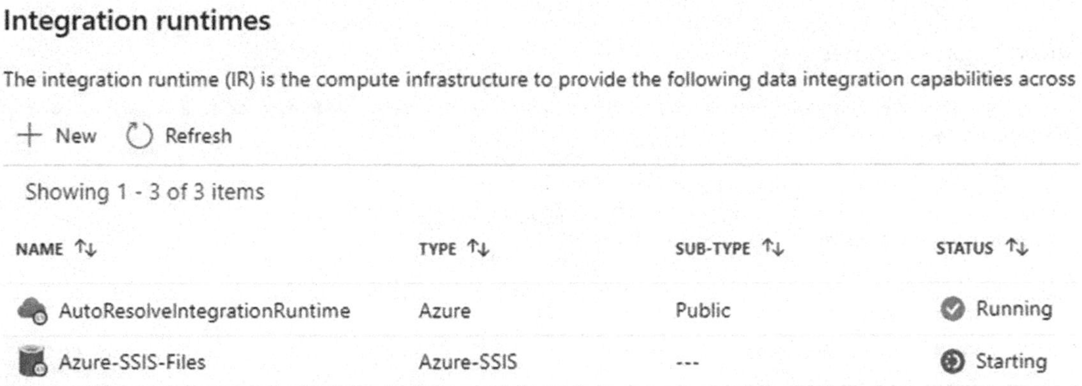

Integration runtimes

The integration runtime (IR) is the compute infrastructure to provide the following data integration capabilities across

+ New   ○ Refresh

Showing 1 - 3 of 3 items

NAME ↑↓	TYPE ↑↓	SUB-TYPE ↑↓	STATUS ↑↓
AutoResolveIntegrationRuntime	Azure	Public	✓ Running
Azure-SSIS-Files	Azure-SSIS	---	⊙ Starting

Figure 8-39. *Azure-SSIS-Files is starting*

After some time – usually 3–5 minutes (maximum) at the time of this writing – the Azure-SSIS IR is started and Connections ➤ Integration runtimes displays "Running" status, as shown in Figure 8-40.

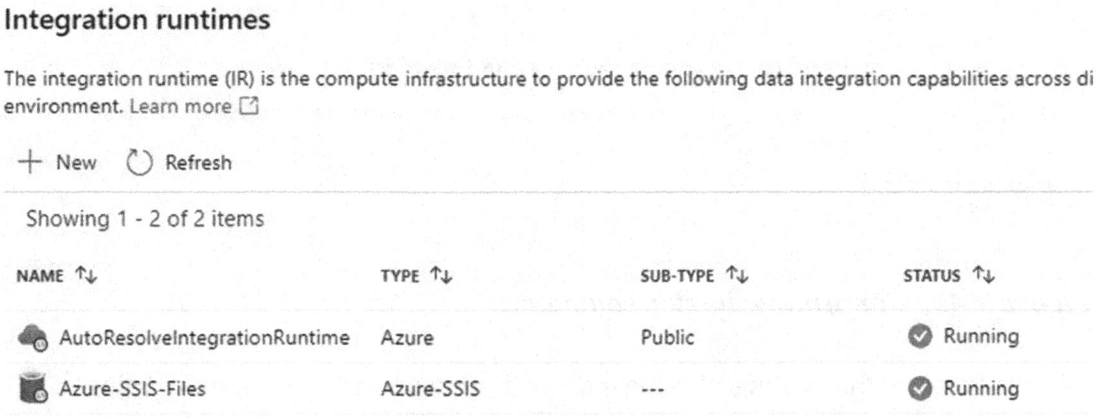

Integration runtimes

The integration runtime (IR) is the compute infrastructure to provide the following data integration capabilities across di environment. Learn more ☐

+ New   ○ Refresh

Showing 1 - 2 of 2 items

NAME ↑↓	TYPE ↑↓	SUB-TYPE ↑↓	STATUS ↑↓
AutoResolveIntegrationRuntime	Azure	Public	✓ Running
Azure-SSIS-Files	Azure-SSIS	---	✓ Running

Figure 8-40. *Azure-SSIS-Files has started*

Stopping the Azure-SSIS Integration Runtime

One way to stop an Azure-SSIS IR is using the "Stop" button (the Stop button displays a Pause icon) found on Connections ➤ Integration runtimes, as shown in Figure 8-41.

Integration runtimes

The integration runtime (IR) is the compute infrastructure to provide the following data integration capabilities across different environment. Learn more ⬈

+ New   ↻ Refresh

Showing 1 - 2 of 2 items

NAME ↑↓	TYPE ↑↓	SUB-TYPE ↑↓	STATUS ↑↓
AutoResolveIntegrationRuntime	Azure	Public	✓ Running
Azure-SSIS-Files ⊘ ⏸ ⟲ ⋯	Azure-SSIS	---	✓ Running

Figure 8-41.   Stopping the Azure-SSIS IR

Clicking the Stop button triggers a confirmation dialog, as shown in Figure 8-42.

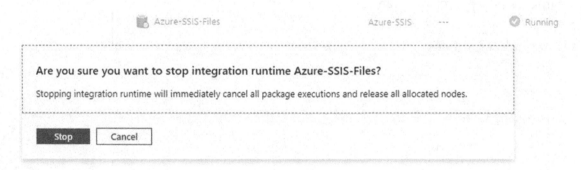

Figure 8-42.   Confirming the stop command

At the time of this writing, clicking the Stop button on the confirmation dialog triggers the survey form shown in Figure 8-43.

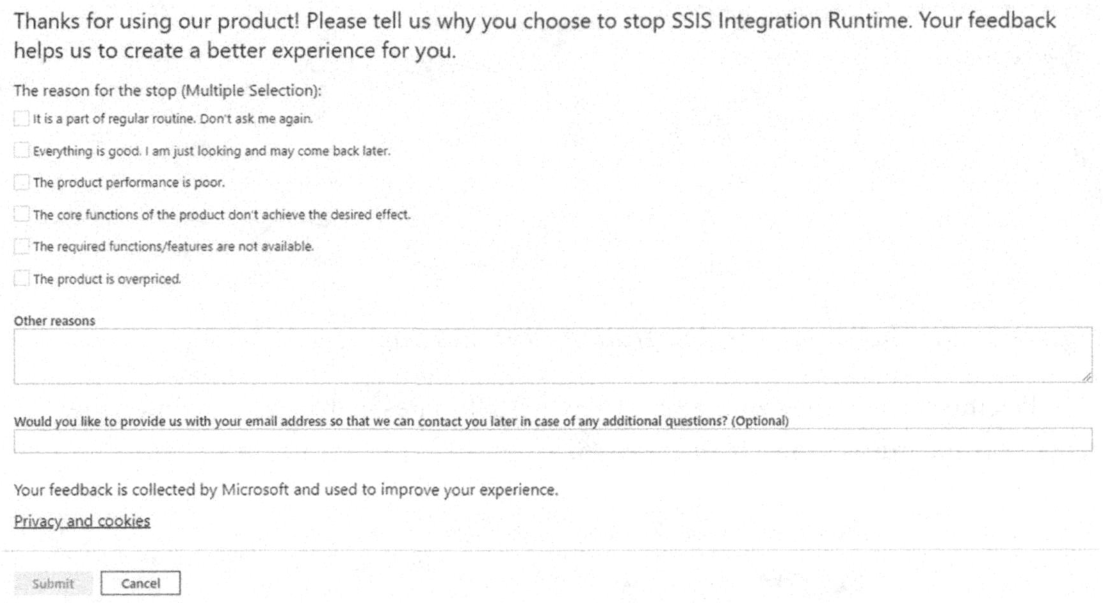

Thanks for using our product! Please tell us why you choose to stop SSIS Integration Runtime. Your feedback helps us to create a better experience for you.

The reason for the stop (Multiple Selection):

☐ It is a part of regular routine. Don't ask me again.

☐ Everything is good. I am just looking and may come back later.

☐ The product performance is poor.

☐ The core functions of the product don't achieve the desired effect.

☐ The required functions/features are not available.

☐ The product is overpriced.

Other reasons

Would you like to provide us with your email address so that we can contact you later in case of any additional questions? (Optional)

Your feedback is collected by Microsoft and used to improve your experience.

Privacy and cookies

Submit   Cancel

Figure 8-43. *A survey dialog triggered by the confirmation dialog*

Once the survey form is submitted (or cancelled), the Azure-SSIS integration runtime enters a "Stopping" state, as shown in Figure 8-44.

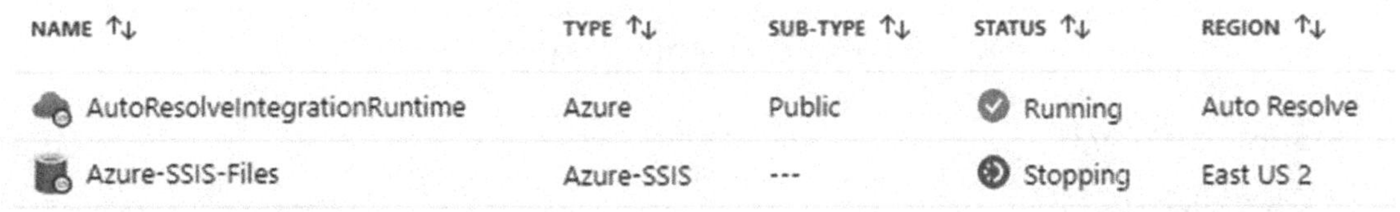

NAME ↑↓	TYPE ↑↓	SUB-TYPE ↑↓	STATUS ↑↓	REGION ↑↓
AutoResolveIntegrationRuntime	Azure	Public	✅ Running	Auto Resolve
Azure-SSIS-Files	Azure-SSIS	---	🔵 Stopping	East US 2

Figure 8-44. *Azure-SSIS is stopping*

Another way to stop the Azure-SSIS IR is from the Monitor ➤ Integration runtimes page, as shown in Figure 8-45.

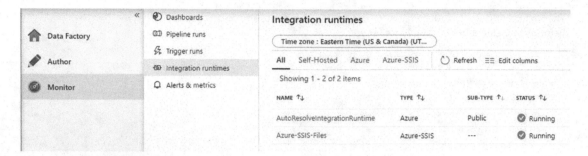

Figure 8-45. *The Monitor ➤ Integration runtimes page*

Click the name of the Azure-SSIS IR – Azure-SSIS-Files in this case – to open the details dashboard, as shown in Figure 8-46.

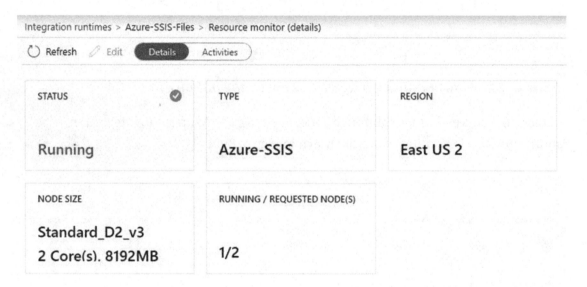

Figure 8-46. *Azure-SSIS details dashboard*

Click the Status ("Running") to open the Azure-SSIS status dialog shown in Figure 8-47.

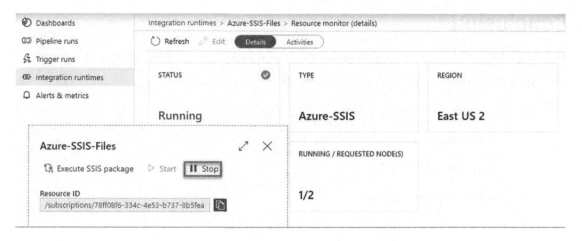

Figure 8-47. *Azure-SSIS status dialog*

Click the Stop button to stop the Azure-SSIS integration runtime, as shown in Figure 8-48.

Figure 8-48. *Azure-SSIS integration runtime stopping*

Please note, clicking the Stop button on the Azure-SSIS status dialog triggers the "Are you sure?/Survey" process.

Restart the Azure-SSIS integration runtime to continue.

Now that an Azure-SSIS IR has been configured for execution from Azure File Shares, the next step is to provision an Azure SQL database.

Conclusion

The focus of this chapter has been provisioning an instance of Azure-SSIS and prerequisites. This chapter walked through the following steps:

- Getting started with Azure

- Provisioning an Azure Data Factory

- Provisioning Azure Storage

- Provisioning Azure-SSIS

The next step is to provision an Azure SQL Database.

CHAPTER 9

Deploy a Simple, Custom, File-Based Azure-SSIS Framework

The Microsoft Azure-SSIS team has been hard at work reducing the friction between executing SSIS packages on-premises and executing SSIS packages in Azure. Many enterprises execute SSIS packages stored in the file system on-premises. In the past, migrating SSIS packages executed in an on-premises file system meant changing enterprise SSIS storage and execution patterns. With the advent of Azure-SSIS File Share–based execution, executing SSIS packages from files in Azure is no longer an issue.

In fact, the SSIS framework in previous chapters works well in Azure-SSIS. In this chapter, we will discuss and demonstrate the following.

Provisioning the SSISConfig Database

The focus of this section is provisioning the SSISConfig database to an instance of Azure SQL Database upon which the SSISConfig database will reside.

Azure SQL can be a somewhat loose term. It can be taken to mean Azure SQL Database, Azure SQL Managed Instance, and some even use it to mean SQL Server running on an Azure VM. In this chapter, we are working in Azure SQL Database, which is the Software as a Service offering as described at docs.microsoft.com/en-us/azure/azure-sql/azure-sql-iaas-vs-paas-what-is-overview.

© Andy Leonard, Kent Bradshaw 2020
A. Leonard and K. Bradshaw, *SQL Server Data Automation Through Frameworks*,
https://doi.org/10.1007/978-1-4842-6213-9_9

Navigate to the Azure portal to begin the provisioning process. In the left Azure menu, hover over "SQL databases" until the SQL databases hover card displays. When the SQL databases hover card displays, click "+ Create" similar to Figure 9-1.

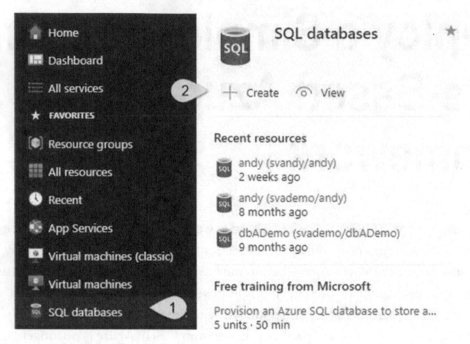

Figure 9-1. *Starting the Azure SQL Database provisioning process*

Clicking "+ Create" opens the Create SQL Database blade. Begin by configuring the Subscription property to your subscription name and the Resource group property to your resource group name, as shown in Figure 9-2.

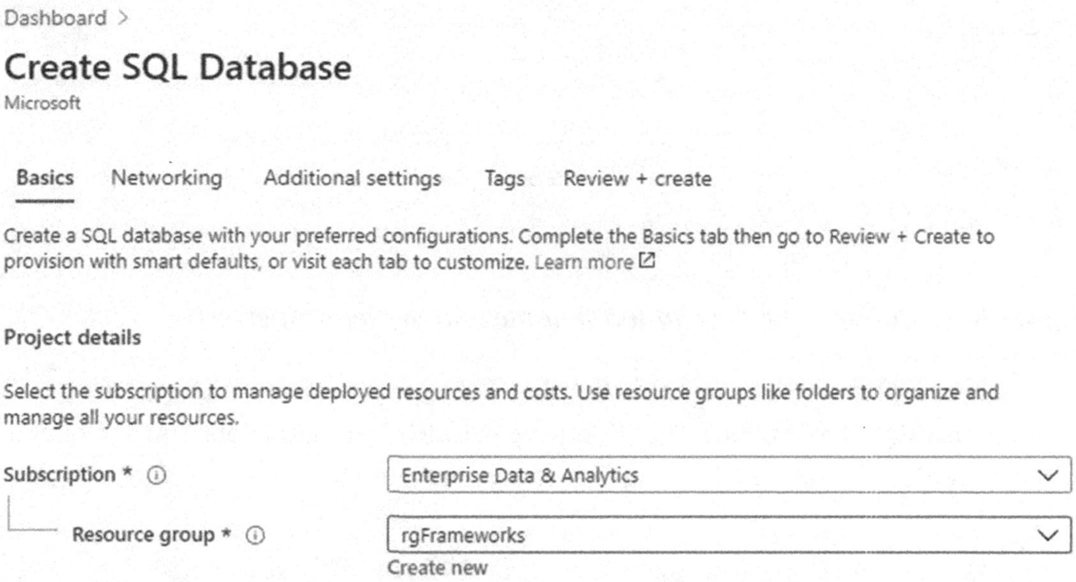

Dashboard >

Create SQL Database

Microsoft

Basics   Networking   Additional settings   Tags   Review + create

Create a SQL database with your preferred configurations. Complete the Basics tab then go to Review + Create to provision with smart defaults, or visit each tab to customize. Learn more ☑

Project details

Select the subscription to manage deployed resources and costs. Use resource groups like folders to organize and manage all your resources.

Subscription * ⓘ	Enterprise Data & Analytics ∨
Resource group * ⓘ	rgFrameworks ∨
	Create new

Figure 9-2. *Azure SQL Database Subscriptions and Resource group properties configuration*

The example uses the resource group named "rgFrameworks." Azure resource group names are globally unique (and case insensitive), so you will need to use a different name for your resource group.

Next, configure the Data details section by entering "SSISConfig" for the "Database name" property. Please note server names are globally unique, but database names are only unique to the server, as shown in Figure 9-3.

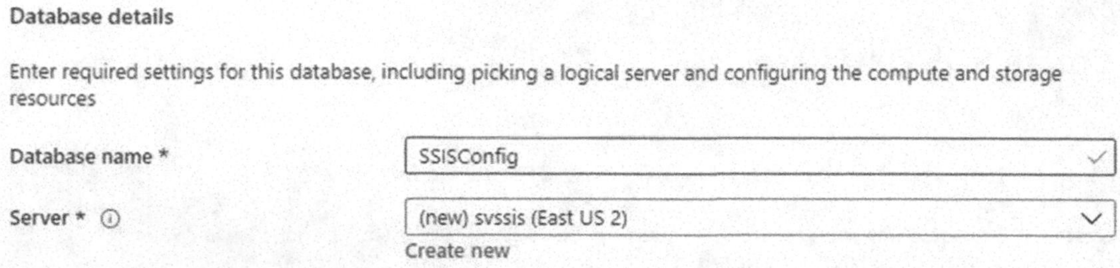

Database details

Enter required settings for this database, including picking a logical server and configuring the compute and storage resources

Database name *

SSISConfig ✓

Server * ⓘ

(new) svssis (East US 2) ⌄
Create new

Figure 9-3. *Configuring the Database name and Server properties*

For the purposes of this example, click the "No" option to answer the question, "Want to use SQL elastic pool?" The "Compute + storage" property defaults to "General purpose Gen5, 2 vCores, 32 GB storage," as shown in Figure 9-4.

Figure 9-4. *SQL elastic pool and "Compute + storage" property configuration*

The default database configuration is a little heavy for the purposes of this example. Click the "Configure database" link shown in Figure 9-4 to open the Configure blade, which will allow changes to database capacity and performance options, as shown in Figure 9-5.

Configure

Figure 9-5. *Configuring the Azure SQL database*

As you can tell from Figure 9-5, the default database configuration is also a little *expensive* for the purposes of this example. Click the "Looking for basic, standard, premium?" link to navigate to, and select, the "Basic" option, as shown in Figure 9-6.

Configure

♡ Feedback

Basic	Standard
For less demanding workloads	For workloads with typical performance requirements

DTUs What is a DTU? ☑

5 (Basic)

Data max size

```
100 MB                                          2 GB
```
[2 GB]

Cost summary

Cost per **DTU** (in USD)	1.00
DTUs selected	x 5
ESTIMATED COST / MONTH	4.99 USD

[Apply]

Figure 9-6. Reconfiguring the Azure SQL database

Two GB and five DTUs is plenty for the example. Click the Apply button to proceed. Database configuration now displays out updated settings, as shown in Figure 9-7.

Want to use SQL elastic pool? *
Yes
No
Compute + storage *
Basic
2 GB storage
Configure database

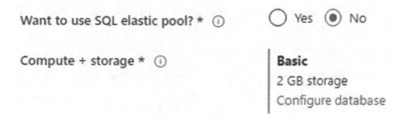

Figure 9-7. *Updated Compute + storage settings*

Before creating the new SSISConfig database, double-check the "Additional settings" tab's "Enable advanced data security" property, as shown in Figure 9-8.

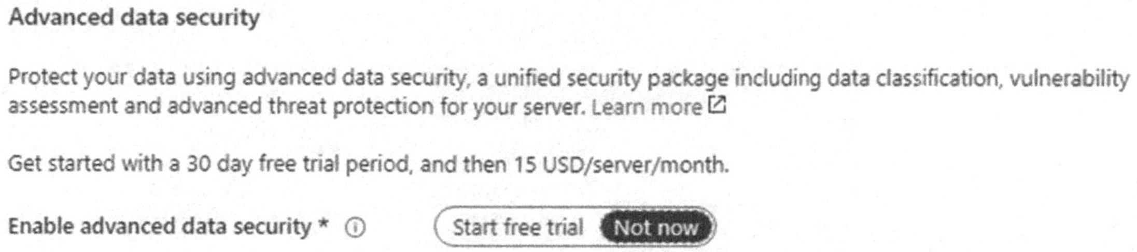
Advanced data security
Protect your data using advanced data security, a unified security package including data classification, vulnerability assessment and advanced threat protection for your server. Learn more
Get started with a 30 day free trial period, and then 15 USD/server/month.
Enable advanced data security *
Start free trial
Not now

Figure 9-8. *Configuring the "Enable advanced data security" property*

If the "Start free trial" option is selected, change the setting to "Not now" for the purposes of this example.

On the Networking and Tags tabs, leave the default configuration settings.

The "Review + create" tab should appear similar to Figure 9-9.

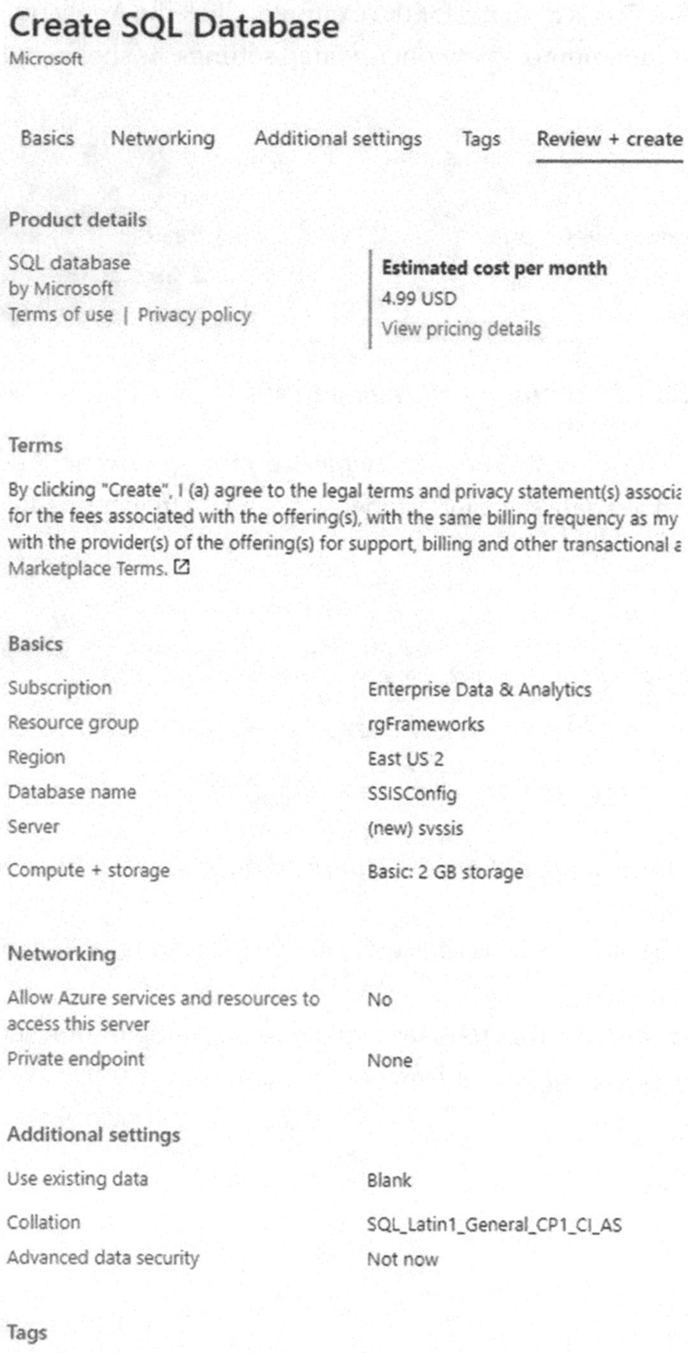

Create SQL Database

Microsoft

Basics   Networking   Additional settings   Tags   **Review + create**

Product details

SQL database
by Microsoft
Terms of use | Privacy policy

| **Estimated cost per month**
| 4.99 USD
| View pricing details

Terms

By clicking "Create", I (a) agree to the legal terms and privacy statement(s) associ:
for the fees associated with the offering(s), with the same billing frequency as my
with the provider(s) of the offering(s) for support, billing and other transactional a
Marketplace Terms. ☑

Basics

Subscription   Enterprise Data & Analytics

Resource group   rgFrameworks

Region   East US 2

Database name   SSISConfig

Server   (new) svssis

Compute + storage   Basic: 2 GB storage

Networking

Allow Azure services and resources to   No
access this server

Private endpoint   None

Additional settings

Use existing data   Blank

Collation   SQL_Latin1_General_CP1_CI_AS

Advanced data security   Not now

Tags

Figure 9-9. *Azure SQL Database Review + create page*

Click the Create button to create the Azure SQL Database. Provisioning an Azure SQL Database takes a few minutes. Once complete, the portal should appear similar to Figure 9-10.

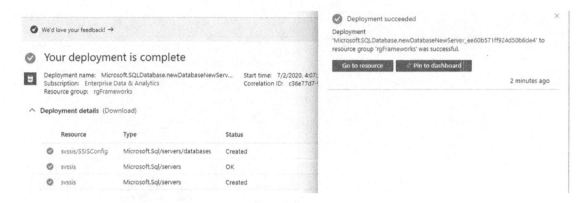

Figure 9-10. *Azure SQL Database, provisioned*

The next step is adding SSISConfig database artifacts to the new Azure SQL Database.

Deploy the Simple, Custom, File-Based Azure-SSIS Framework

The focus of this section is deploying the SSISConfig database designed in Chapters 5 and 7 to an Azure SQL SSISConfig database instance.

To begin, open Azure Data Studio (or SQL Server Management Studio) and connect to the recently provisioned Azure SQL Database instance, as shown in Figure 9-11.

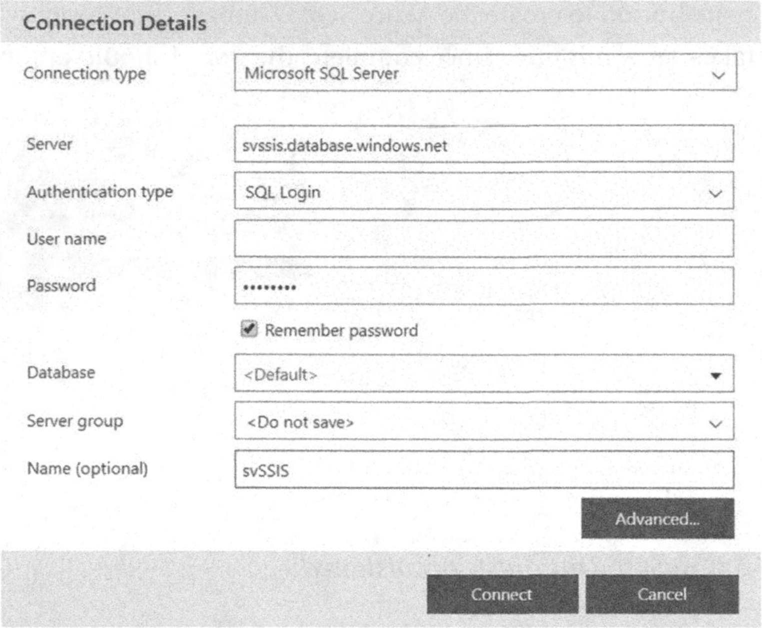

Connection Details

Connection type	Microsoft SQL Server ⌄
Server	svssis.database.windows.net
Authentication type	SQL Login ⌄
User name	
Password	••••••••
	☑ Remember password
Database	\<Default\> ▾
Server group	\<Do not save\> ⌄
Name (optional)	svSSIS
	Advanced...

Connect   Cancel

Figure 9-11. Connecting to the new Azure SQL Database

The SSISConfig database is created in the previous step. A script to create the SSISConfig database is similar to the scripts used in Chapters 5 and 7, as shown in Listings 9-1.

Listing 9-1. Creating SSISConfig in the Azure SQL database

```
print 'SSISConfig database'
If Not Exists(Select [databases].[name]
            From [sys].[databases]
            Where [databases].[name] = N'SSISConfig')
 begin
  print ' - Create SSISConfig database'
  Create Database SSISConfig
  print ' - SSISConfig database created'
 end
Else
```

```
begin
  print ' - SSISConfig database already exists.'
 end
print ''
go
```

When execution completes, Azure Data Studio should display messages similar to the messages shown in Figure 9-12.

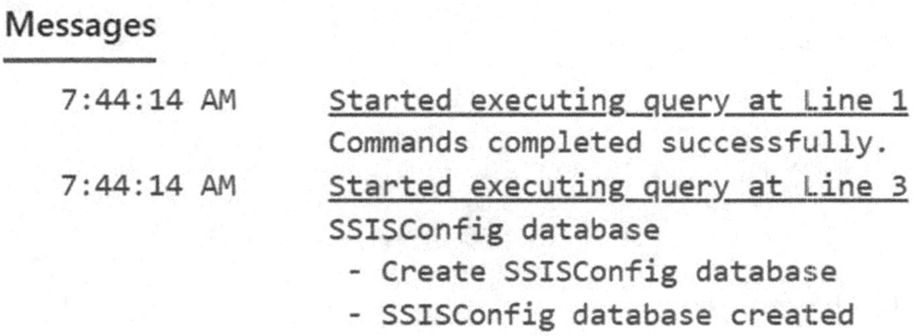

Messages

7:44:14 AM	Started executing query at Line 1
	Commands completed successfully.
7:44:14 AM	Started executing query at Line 3
	SSISConfig database
	- Create SSISConfig database
	- SSISConfig database created

Figure 9-12. *SSISConfig database created*

At the time of this writing, the T-SQL in Azure SQL databases does not support the Use command. If you attempt to use the Use command, the error shown in Figure 9-13 is generated.

Instead of the Use command, use the database selector drop-down highlighted in Figure 9-13.

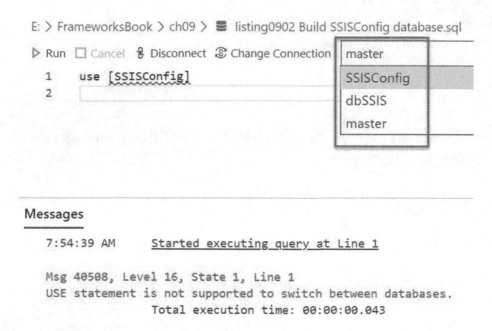

Figure 9-13. *It's no "use"...*

Listing 9-2 combines the remaining SSISConfig DDL (data definition language) statements found in Chapters 5 and 7 to create the SSISConfig database artifacts and metadata.

Listing 9-2. Creating SSISConfig database artifacts

```
print 'Config schema'
If Not Exists(Select [schemas].[name]
              From [sys].[schemas]
              Where [schemas].[name] = N'config')
 begin
  print ' - Create config schema'
  declare @sql nvarchar(100) = N'Create Schema config'
  exec(@sql)
  print ' - Config schema created'
 end
Else
```

```
 begin
  print ' - Config schema already exists.'
 end
print ''
go

print 'Config.Applications table'
If Not Exists(Select [schemas].[name] + '.' + [tables].[name]
As [Schema.Table]
             From [sys].[tables]
             Join [sys].[schemas]
               On [schemas].[schema_id] = [tables].[schema_id]
             Where [schemas].[name] = N'config'
               And [tables].[name] = N'Applications')
 begin
  print ' - Create config.Applications table'
  Create Table [config].[Applications]
   (
      ApplicationId int identity(1, 1)
        Constraint PK_config_Applications Primary Key Clustered
    , ApplicationName nvarchar(255) Not NULL
        Constraint UQ_config_Applications_ApplicationName
         Unique
   )
  print ' - Config.Applications table created'
 end
Else
 begin
  print ' - Config.Applications table already exists.'
 end
print ''
go
```

```
print 'Config.Packages table'
If Not Exists(Select [schemas].[name] + '.' + [tables].[name]
As [Schema.Table]
             From [sys].[tables]
             Join [sys].[schemas]
               On [schemas].[schema_id] = [tables].[schema_id]
             Where [schemas].[name] = N'config'
               And [tables].[name] = N'Packages')
 begin
  print ' - Create config.Packages table'
  Create Table [config].[Packages]
   (
      PackageId int identity(1, 1)
       Constraint PK_config_Packages Primary Key Clustered
    , PackageLocation nvarchar(255) Not NULL
    , PackageName nvarchar(255) Not NULL
    , Constraint UQ_config_Packages_PackageName
       Unique(PackageLocation, PackageName)
   )
  print ' - Config.Packages table created'
 end
Else
 begin
  print ' - Config.Packages table already exists.'
 end
print ''
go

print 'Config.ApplicationPackages table'
If Not Exists(Select [schemas].[name] + '.' + [tables].[name]
As [Schema.Table]
             From [sys].[tables]
             Join [sys].[schemas]
               On [schemas].[schema_id] = [tables].[schema_id]
             Where [schemas].[name] = N'config'
               And [tables].[name] = N'ApplicationPackages')
```

```
begin
 print ' - Create config.ApplicationPackages table'
 Create Table [config].[ApplicationPackages]
  (
     ApplicationPackageId int identity(1, 1)
       Constraint PK_config_ApplicationPackages Primary Key Clustered
   , ApplicationId int Not NULL
       Constraint FK_config_ApplicationPackages_config_Applications
         Foreign Key References [config].[Applications](ApplicationId)
   , PackageId int Not NULL
       Constraint FK_config_ApplicationPackages_config_Packages
         Foreign Key References [config].[Packages](PackageId)
   , ExecutionOrder int Not NULL
       Constraint DF_config_ApplicationPackages_ExecutionOrder
         Default(10)
   , ApplicationPackageEnabled bit Not NULL
       Constraint DF_config_ApplicationPackages_ApplicationPackageEnabled
         Default(1)
   , FailApplicationOnPackageFailure bit Not NULL
       Constraint DF_config_ApplicationPackages_
       FailApplicationOnPackageFailure
         Default(1)
   , Constraint UQ_config_ApplicationPackages_ApplicationId_PackageId_
     ExecutionOrder
       Unique(ApplicationId, PackageId, ExecutionOrder)
  )
 print ' - Config.ApplicationPackages table created'
 end
Else
 begin
 print ' - Config.ApplicationPackages table already exists.'
 end
print ''
go
```

```
Set NoCount ON

declare @ApplicationName nvarchar(255) = N'Framework Test'

print @ApplicationName
declare @ApplicationId int = (Select [Applications].[ApplicationId]
                             From [config].[Applications]
                             Where [Applications].[ApplicationName] =
                             @ApplicationName)
If (@ApplicationId Is NULL)
 begin
  print ' - Adding ' + @ApplicationName + ' application to config.
  Applications table'

  declare @AppTbl table(ApplicationId int)

  Insert Into [config].[Applications]
  (ApplicationName)
  Output inserted.ApplicationId into @AppTbl
  Values (@ApplicationName)

  Set @ApplicationId = (Select ApplicationId
                        From @AppTbl)

  print ' - ' + @ApplicationName + ' application added to config.
  Applications table'
 end
Else
 begin
  print ' - ' + @ApplicationName + ' application already exists in the
  config.Applications table.'
 end

 Select @ApplicationId As ApplicationId
print ''
go

Set NoCount ON
```

```
declare @PackageLocation nvarchar(255) = N'E:\Projects\TestSSISSolution\
TestSSISProject\'
declare @PackageName nvarchar(255) = N'ReportAndSucceed.dtsx'

print @PackageLocation + @PackageName
declare @PackageId int = (Select [Packages].[PackageId]
                          From [config].[Packages]
                          Where [Packages].[PackageLocation] = @
PackageLocation
                          And [Packages].[PackageName] = @PackageName)
If (@PackageId Is NULL)
 begin
  print ' - Adding ' + @PackageName + ' package to config.Packages table'

  declare @PkgTbl table(PackageId int)

  Insert Into [config].[Packages]
  (PackageLocation, PackageName)
  Output inserted.PackageId into @PkgTbl
  Values (@PackageLocation, @PackageName)

  Set @PackageId = (Select PackageId
                    From @PkgTbl)

  print ' - ' + @PackageName + ' package added to config.Packages table'
 end
Else
 begin
  print ' - ' + @PackageName + ' package already exists in the config.
  Packages table.'
 end

 Select @PackageId As PackageId
print ''
```

```
set @PackageName = N'ReportAndFail.dtsx'

print @PackageLocation + @PackageName
set @PackageId = (Select [Packages].[PackageId]
                  From [config].[Packages]
                  Where [Packages].[PackageLocation] = @PackageLocation
                    And [Packages].[PackageName] = @PackageName)
If (@PackageId Is NULL)
 begin
  print ' - Adding ' + @PackageName + ' application to config.Packages
  table'

  Delete @PkgTbl

  Insert Into [config].[Packages]
  (PackageLocation, PackageName)
  Output inserted.PackageId into @PkgTbl
  Values (@PackageLocation, @PackageName)

  Set @PackageId = (Select PackageId
                    From @PkgTbl)

  print ' - ' + @PackageName + ' package added to config.Packages table'
 end
Else
 begin
  print ' - ' + @PackageName + ' package already exists in the config.
  Packages table.'
 end

 Select @PackageId As PackageId
print ''
go

Set NoCount ON

declare @ApplicationName nvarchar(255) = N'Framework Test'
declare @PackageLocation nvarchar(255) = N'E:\Projects\TestSSISSolution\
TestSSISProject\'
declare @PackageName nvarchar(255) = N'ReportAndSucceed.dtsx'
```

258

```
declare @ExecutionOrder int = 10

print @ApplicationName + ' - ' + @PackageLocation + @PackageName

declare @ApplicationId int = (Select [Applications].[ApplicationId]
                              From [config].[Applications]
                              Where [Applications].[ApplicationName] =
                              @ApplicationName)

declare @PackageId int = (Select [Packages].[PackageId]
                          From [config].[Packages]
                          Where [Packages].[PackageLocation] =
                          @PackageLocation
                            And [Packages].[PackageName] = @PackageName)

declare @ApplicationPackageId int = (Select ApplicationPackageId
                                     From config.ApplicationPackages
                                     Where ApplicationId = @ApplicationId
                                       And PackageId = @PackageId
                                       And ExecutionOrder = @
ExecutionOrder)

If (@ApplicationPackageId Is NULL)
 begin
  print ' - Assigning ' + @PackageName + ' package to '
       + @ApplicationName + ' application'
       + ' in config.ApplicationPackages table'
       + ' at ExecutionOrder ' + Convert(varchar(9), @ExecutionOrder)

  Insert Into [config].[ApplicationPackages]
  (ApplicationId
, PackageId
, ExecutionOrder)
  Values (@ApplicationId
       , @PackageId
       , @ExecutionOrder)
```

```
   print ' - ' + @PackageName + ' package assigned to '
        + @ApplicationName + ' application'
        + ' in config.ApplicationPackages table'
        + ' at ExecutionOrder ' + Convert(varchar(9), @ExecutionOrder)
 end
Else
 begin
print ' - ' + @PackageName + ' package already'
        + ' assigned to ' + @ApplicationName
        + ' application in config.ApplicationPackages table'
        + ' at ExecutionOrder ' + Convert(varchar(9), @ExecutionOrder)
        + '.'
 end
print ''

set @PackageName = N'ReportAndFail.dtsx'
set @ExecutionOrder = 20

print @ApplicationName + ' - ' + @PackageLocation + @PackageName

set @ApplicationId = (Select [Applications].[ApplicationId]
                      From [config].[Applications]
                      Where [Applications].[ApplicationName] =
                      @ApplicationName)

set @PackageId = (Select [Packages].[PackageId]
                  From [config].[Packages]
                  Where [Packages].[PackageLocation] = @PackageLocation
                    And [Packages].[PackageName] = @PackageName)

set @ApplicationPackageId = (Select ApplicationPackageId
                             From config.ApplicationPackages
                             Where ApplicationId = @ApplicationId
                               And PackageId = @PackageId
                               And ExecutionOrder = @ExecutionOrder)
```

```
If (@ApplicationPackageId Is NULL)
 begin
   print ' - Assigning ' + @PackageName + ' package to '
        + @ApplicationName + ' application'
        + ' in config.ApplicationPackages table'
        + ' at ExecutionOrder ' + Convert(varchar(9), @ExecutionOrder)

   Insert Into [config].[ApplicationPackages]
   (ApplicationId
 , PackageId
 , ExecutionOrder)
   Values (@ApplicationId
        , @PackageId
        , @ExecutionOrder)

   print ' - ' + @PackageName + ' package assigned to '
        + @ApplicationName + ' application'
        + ' in config.ApplicationPackages table'
        + ' at ExecutionOrder ' + Convert(varchar(9), @ExecutionOrder)
 end
Else
 begin
   print ' - ' + @PackageName + ' package already'
        + ' assigned to ' + @ApplicationName
        + ' application in config.ApplicationPackages table'
        + ' at ExecutionOrder ' + Convert(varchar(9), @ExecutionOrder)
        + '.'
 end
print ''
go

print 'Log schema'
If Not Exists(Select [schemas].[name]
            From [sys].[schemas]
            Where [schemas].[name] = N'log')
```

```
begin
 print ' - Create log schema'
 declare @sql nvarchar(100) = N'Create Schema log'
 exec(@sql)
 print ' - Log schema created'
end
Else
begin
 print ' - Log schema already exists.'
end
print ''
go

print 'Log.ApplicationInstance table'
If Not Exists(Select [schemas].[name] + '.' + [tables].[name]
As [Schema.Table]
             From [sys].[tables]
             Join [sys].[schemas]
               On [schemas].[schema_id] = [tables].[schema_id]
             Where [schemas].[name] = N'log'
                And [tables].[name] = N'ApplicationInstance')
 begin
  print ' - Create log.ApplicationInstance table'
  Create Table [log].[ApplicationInstance]
   (
     ApplicationInstanceId int identity(1, 1)
        Constraint PK_log_ApplicationInstance Primary Key Clustered
    , ApplicationId int Not NULL
        Constraint FK_log_ApplicationInstance_config_Applications
          Foreign Key References [config].[Applications](ApplicationId)
    , ApplicationStartTime datetimeoffset(7) Not NULL
       Constraint DF_log_ApplicationInstance_ApplicationStartTime
         Default(sysdatetimeoffset())
    , ApplicationEndTime datetimeoffset(7) NULL
    , ApplicationStatus nvarchar(25) Not NULL
       Constraint DF_log_ApplicationInstance_ApplicationStatus
```

```
        Default(N'Running')
   )
  print ' - Log.ApplicationInstance table created'
 end
Else
 begin
  print ' - Log.ApplicationInstance table already exists.'
 end
print ''
go

print 'Log.ApplicationPackageInstance table'
If Not Exists(Select [schemas].[name] + '.' + [tables].[name]
As [Schema.Table]
              From [sys].[tables]
              Join [sys].[schemas]
                On [schemas].[schema_id] = [tables].[schema_id]
              Where [schemas].[name] = N'log'
                And [tables].[name] = N'ApplicationPackageInstance')
 begin
  print ' - Create log.ApplicationPackageInstance table'
  Create Table [log].[ApplicationPackageInstance]
  (
     ApplicationPackageInstanceId int identity(1, 1)
        Constraint PK_log_ApplicationPackageInstance Primary Key Clustered
   , ApplicationInstanceId int Not NULL
       Constraint FK_log_ApplicationPackageInstance_log_ApplicationInstance
        Foreign Key References [log].[ApplicationInstance]
        (ApplicationInstanceId)
   , ApplicationPackageId int Not NULL
       Constraint FK_log_ApplicationPackageInstance_config_
       ApplicationPackages
        Foreign Key References [config].[ApplicationPackages]
        (ApplicationPackageId)
   , ApplicationPackageStartTime datetimeoffset(7) Not NULL
```

```
      Constraint DF_log_ApplicationPackageInstance_
      ApplicationPackageStartTime
       Default(sysdatetimeoffset())
    , ApplicationPackageEndTime datetimeoffset(7) NULL
    , ApplicationPackageStatus nvarchar(25) Not NULL
      Constraint DF_log_ApplicationPackageInstance_
      ApplicationPackageStatus
       Default(N'Running')
   )
  print ' - Log.ApplicationPackageInstance table created'
 end
Else
 begin
  print ' - Log.ApplicationPackageInstance table already exists.'
 end
print ''
go
```

If all goes as planned, the script returns results shown in Figure 9-14.

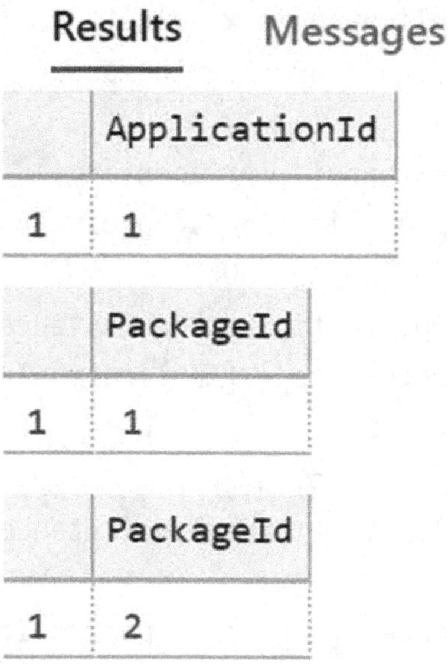

Results   Messages

	ApplicationId
1	1

	PackageId
1	1

	PackageId
1	2

Figure 9-14. *Results of query execution*

The Messages returned from executing the T-SQL query in Listing 9-2 are shown in Listing 9-3.

Listing 9-3. Messages returned from SSISConfig artifact and metadata creation

```
Started executing query at Line 1
Config schema
 - Create config schema
 - Config schema created

Started executing query at Line 17
Config.Applications table
 - Create config.Applications table
 - Config.Applications table created

Started executing query at Line 43
Config.Packages table
 - Create config.Packages table
 - Config.Packages table created

Started executing query at Line 70
Config.ApplicationPackages table
 - Create config.ApplicationPackages table
 - Config.ApplicationPackages table created

Started executing query at Line 110
Framework Test
 - Adding Framework Test application to config.Applications table
 - Framework Test application added to config.Applications table

Started executing query at Line 143
E:\Projects\TestSSISSolution\TestSSISProject\ReportAndSucceed.dtsx
 - Adding ReportAndSucceed.dtsx package to config.Packages table
 - ReportAndSucceed.dtsx package added to config.Packages table

E:\Projects\TestSSISSolution\TestSSISProject\ReportAndFail.dtsx
 - Adding ReportAndFail.dtsx application to config.Packages table
 - ReportAndFail.dtsx package added to config.Packages table
```

```
Started executing query at Line 209
Framework Test - E:\Projects\TestSSISSolution\TestSSISProject\
ReportAndSucceed.dtsx
 - Assigning ReportAndSucceed.dtsx package to Framework Test application ➤
   in config.ApplicationPackages table at ExecutionOrder 10
 - ReportAndSucceed.dtsx package assigned to Framework Test application ➤
   in config.ApplicationPackages table at ExecutionOrder 10

Framework Test - E:\Projects\TestSSISSolution\TestSSISProject\
ReportAndFail.dtsx
 - Assigning ReportAndFail.dtsx package to Framework Test application ➤
   in config.ApplicationPackages table at ExecutionOrder 20
 - ReportAndFail.dtsx package assigned to Framework Test application ➤
   in config.ApplicationPackages table at ExecutionOrder 20

Log.ApplicationInstance table
 - Create log.ApplicationInstance table
 - Log.ApplicationInstance table created

Log.ApplicationPackageInstance table
 - Create log.ApplicationPackageInstance table
 - Log.ApplicationPackageInstance table created

Total execution time: 00:00:00.557
```

Test the deployment and test metadata insert using the T-SQL query shown in
Listing 9-4.

Listing 9-4. Retrieve SSIS application metadata

```
Select a.ApplicationName
     , p.PackageLocation + p.PackageName As PackagePath
     , ap.ExecutionOrder
     , ap.FailApplicationOnPackageFailure
From [config].[ApplicationPackages] ap
Join [config].[Applications] a
  On a.ApplicationId = ap.ApplicationId
Join [config].Packages p
  On p.PackageId = ap.PackageId
```

```
Where a.ApplicationName = N'Framework Test'
  And ap.ApplicationPackageEnabled = 1
Order By ap.ExecutionOrder
```

If all has gone according to plan, your results should appear similar to Figure 9-15.

ApplicationName	PackagePath	ExecutionOrder	FailApplicationOnPackageFailure
Framework Test	E:\Projects\TestSSISSolution\TestSSISProject\ReportAndSucceed.dtsx	10	1
Framework Test	E:\Projects\TestSSISSolution\TestSSISProject\ReportAndFail.dtsx	20	1

Figure 9-15. *A Test SSIS application in SSISConfig in Azure SQL database*

If your SSISConfig database query results appear as shown in Figure 9-10, congratulations! You followed the instructions correctly. But there's an issue with the instructions: how will the Azure Data Factory version of the SSIS execution engine find my – or your – local drive (my E drive, in this case)? The short answer is, "That would be difficult."

Before we update the PackagePath metadata, let's first provision an Azure File Share.

Provision an Azure File Share

Begin by downloading Microsoft Azure Storage Explorer at storageexplorer.com, as shown in Figure 9-16.

Figure 9-16. *Browse to StorageExplorer.com to download Azure Storage Explorer*

Install Azure Storage Explorer, and connect to the Azure Storage account provisioned in the previous chapter. When connected, storage explorer surfaces an Explorer window that appears similar to Figure 9-17.

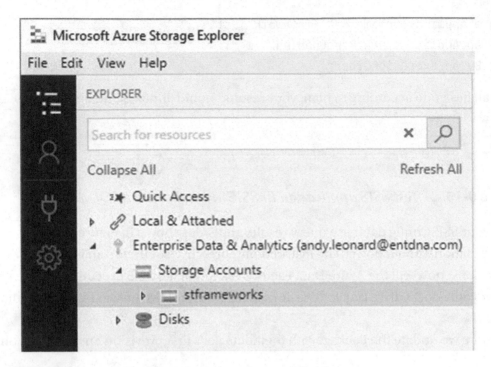

Figure 9-17. *Viewing Explorer in Azure Storage Explorer*

In Chapter 8, we provisioned a storage account named "stframeworks" which we see in the screenshot of Microsoft Azure Storage Explorer in Figure 9-17. Expand the stframeworks storage account to surface virtual folders for Blob Containers, File Shares, Queries, and Tables, as shown in Figure 9-18.

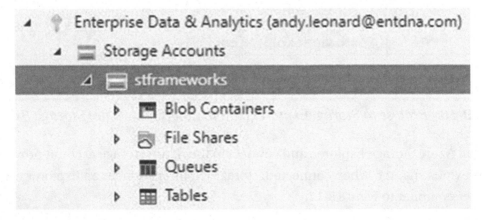

Figure 9-18. *stframeworks virtual folders*

Right-click the File Shares virtual folder, and click "Create File Share" as shown in Figure 9-19.

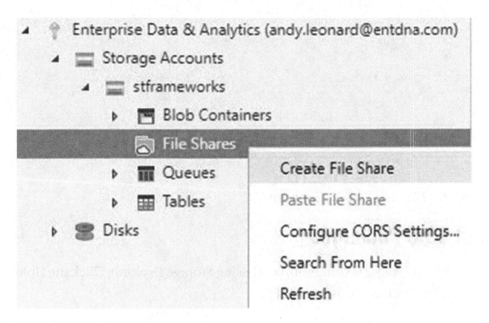

Figure 9-19. *Creating a new File Share*

When the new file share displays, enter a name for the file share – such as "fs-ssis," as shown in Figure 9-20.

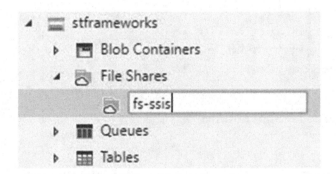

Figure 9-20. *Naming the new file share*

Press the Enter key to finish creating the file share.

Our new file share opens and is now ready to store SSIS packages in Azure, as shown in Figure 9-21.

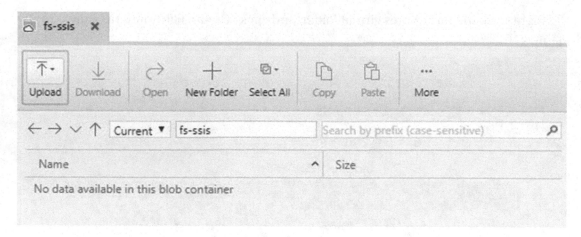

Figure 9-21. *The new fs-ssis file share*

Upload SSIS Packages

Uploading SSIS packages is straightforward using Storage Explorer. Click the Upload button to begin, as shown in Figure 9-22.

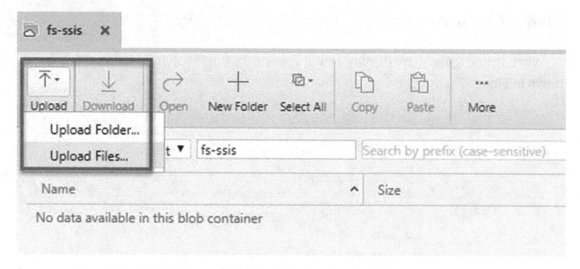

Figure 9-22. *Preparing to upload SSIS packages to the file share*

You may upload a folder or files, as shown in Figure 9-22.

Click "Upload Files..." to open the Upload Files dialog. First, select a file or files to be uploaded – such as the ReportAndSucceed.dtsx and ReportAndFail.dtsx SSIS packages developed as described in Chapter 5. Second, accept the default Destination directory ("/"). Third, click the Upload button, as shown in Figure 9-23.

Figure 9-23. *Configuring the upload to the file share*

When the upload completes, you may view the results of the operation in the Azure Storage Explorer Activities window, as shown in Figure 9-24.

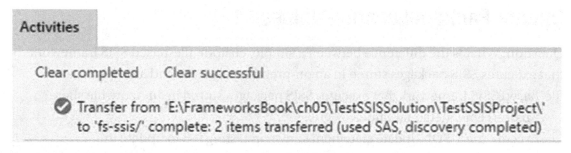

Figure 9-24. *Viewing activities associated with the upload*

ReportAndSucceed.dtsx and ReportAndFail.dtsx SSIS packages have been uploaded and are now visible in the fs-ssis file share, as shown in Figure 9-25.

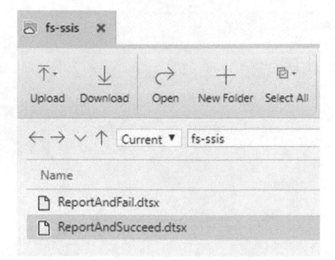

Figure 9-25. *ReportAndSucceed.dtsx and ReportAndFail.dtsx SSIS packages uploaded*

The next step is to update the metadata in the SSISConfig config.Packages table to reflect the file share location of the packages.

Update PackageLocation Values

Question: What is the difference between a simple, custom, file-based SSIS framework that executes SSIS packages stored in an on-premises file store and a simple, custom, file-based SSIS framework that executes SSIS packages stored in an Azure file share?

Answer: One metadata value.

Execute the T-SQL Update statement shown in Listing 9-5 to update the PackageLocation values stored in the config.Packages table.

Listing 9-5. Updating the config.Packages' PackageLocation values

```
Select p.PackageLocation + p.PackageName As PackagePath
From [config].[Packages] p

Update [config].[Packages]
Set PackageLocation  = N'\\stframeworks.file.core.windows.net\fs-ssis\'
Where PackageLocation  = N'E:\Projects\TestSSISSolution\TestSSISProject\'

Select p.PackageLocation + p.PackageName As PackagePath
From [config].[Packages] p
```

Results from executing the T-SQL shown in Listing 9-5 should appear similar to Figure 9-26.

Results   Messages

	PackagePath
1	E:\Projects\TestSSISSolution\TestSSISProject\ReportAndFail.dtsx
2	E:\Projects\TestSSISSolution\TestSSISProject\ReportAndSucceed.dtsx

	PackagePath
1	\\stframeworks.file.core.windows.net\fs-ssis\ReportAndFail.dtsx
2	\\stframeworks.file.core.windows.net\fs-ssis\ReportAndSucceed.dtsx

Figure 9-26. *Updating the config.Packages metadata*

We are now ready to build the SSIS framework execution engine, Azure Data Factory edition.

Build the SSIS Framework ADF Execution Engine

Until this point, deploying a simple, custom, file-based SSIS framework that executes SSIS packages stored in an Azure file share required no changes to the underlying architecture of the SSIS framework metadata database, only metadata was changed, and then only the values stored in a single field.

The SSIS framework execution engine is an Azure Data Factory pipeline, and pipelines are different from SSIS packages. These differences will drive changes in the database and elsewhere.

Retrieve a List of SSIS Packages

To begin building the SSIS framework execution engine in ADF, connect to the Author and Monitor site for the Azure Data Factory instance, click the "+" beside the "Filter resources by name" textbox, and then click "Pipeline" to create a new pipeline, as shown in Figure 9-27.

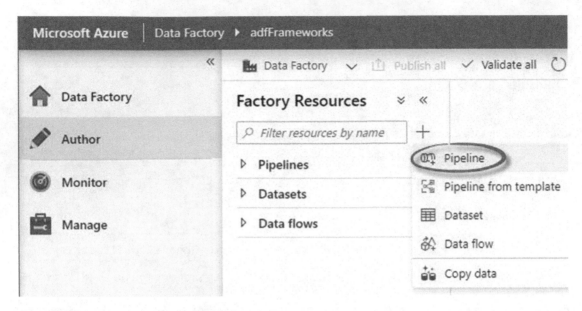

Figure 9-27. *Creating a new ADF pipeline*

When the new pipeline opens, change the name to "parent," as shown in Figure 9-28.

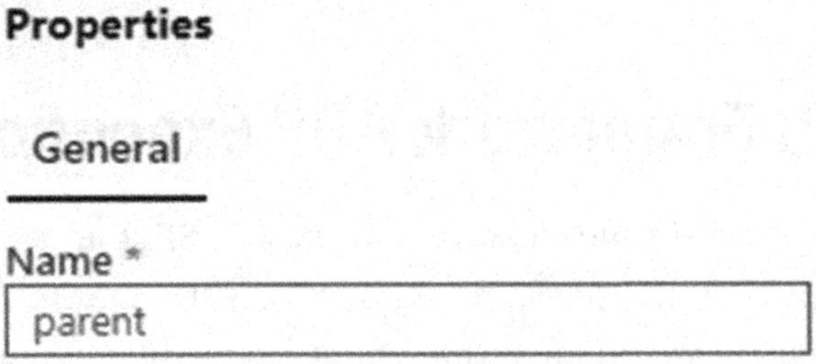

Properties

General
_____

Name *

┌───┐
│ parent   │
└───┘

Figure 9-28. *Renaming the new pipeline "parent"*

In the Activities blade, expand the General category, and then drag a Lookup activity onto the pipeline canvas. Rename the Lookup activity to "Get Application Packages," as shown in Figure 9-29.

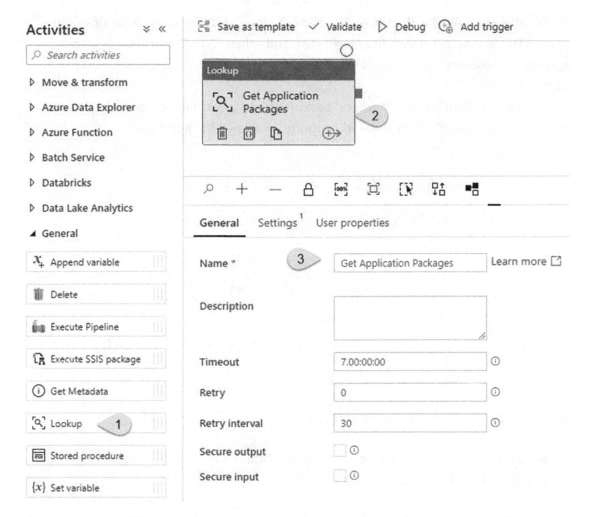

Figure 9-29. *Adding and renaming a Lookup activity to the parent pipeline*

One nice feature of Azure Data Factory is flexible order of authoring, or "FOA." FOA means one may approach ADF development from either direction, bottom up or top down.

Bottom-up ADF development involves building Linked Services *first* – to allow pipeline artifacts to communicate with data stores and services that reside outside the pipeline. Once Linked Services are built in bottom-up fashion, Datasets are developed. Datasets consume Linked Services and surface collections of data to pipeline artifacts. If you are familiar with SSIS, ADF Linked Services are analogous to SSIS Connection Managers, and ADF Datasets are (less) analogous to SSIS Data Flow source and destination adapters.

For this example, I will leverage FOA for top-down development. What does top-down ADF development look like? I will click and select "New" buttons, links, and options to construct ADF artifacts like Datasets and Linked Services in the next few sections.

On the "Get Application Packages" lookup activity's Settings tab, click the "+ New" link to configure the "Source dataset" property, as shown in Figure 9-30.

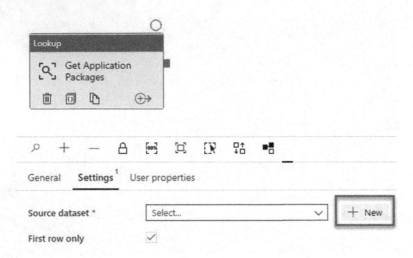

Figure 9-30. *Add a new Source dataset*

When the New dataset blade displays, select "Azure SQL Database," and then click the Continue button, as shown in Figure 9-31.

New dataset

In pipeline activities and data flows, reference a dataset to specify the location and structure of you data within a data store. Learn more 🗗

Select a data store

Figure 9-31. *Configuring the New dataset's data store type*

Set the dataset's Name property; I named my dataset "ssisFrameworkDataSet". Click the Linked Service drop-down, and then click the "+ New" option as shown in Figure 9-32.

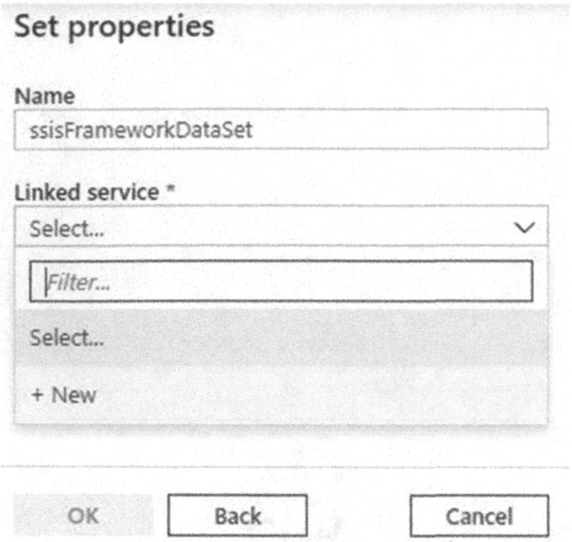

Set properties

Name

ssisFrameworkDataSet

Linked service *

| Select... | ⌄ |

| Filter... |

Select...

+ New

| OK | Back | Cancel |

Figure 9-32.   Configuring the dataset Name and Linked service properties

When the "New linked service (Azure SQL Database)" page displays, enter a Name and optional Description. Leave the "Connect via integration runtime" option set to its default value, which is "AutoResolveIntegrationRuntime".

For the purposes of this example, select the "Connection string" connection source, and then select the "From Azure subscription" option for the "Account selection method" property. Select your "Azure subscription" in the subscription drop-down, select the name of the server you created when provisioning the Azure SQL database from the "Server name" drop-down, and then select the name of the Azure SQL database you provisioned from the "Database name" drop-down.

Select "SQL authentication" in the "Authentication type" drop-down, and enter your Azure SQL database user name and password in the "User name" and "Password" textboxes, respectively.

You are now ready to test the connection. Click the "Test connection" link to test connectivity. The connection test should fail, as shown in Figure 9-33.

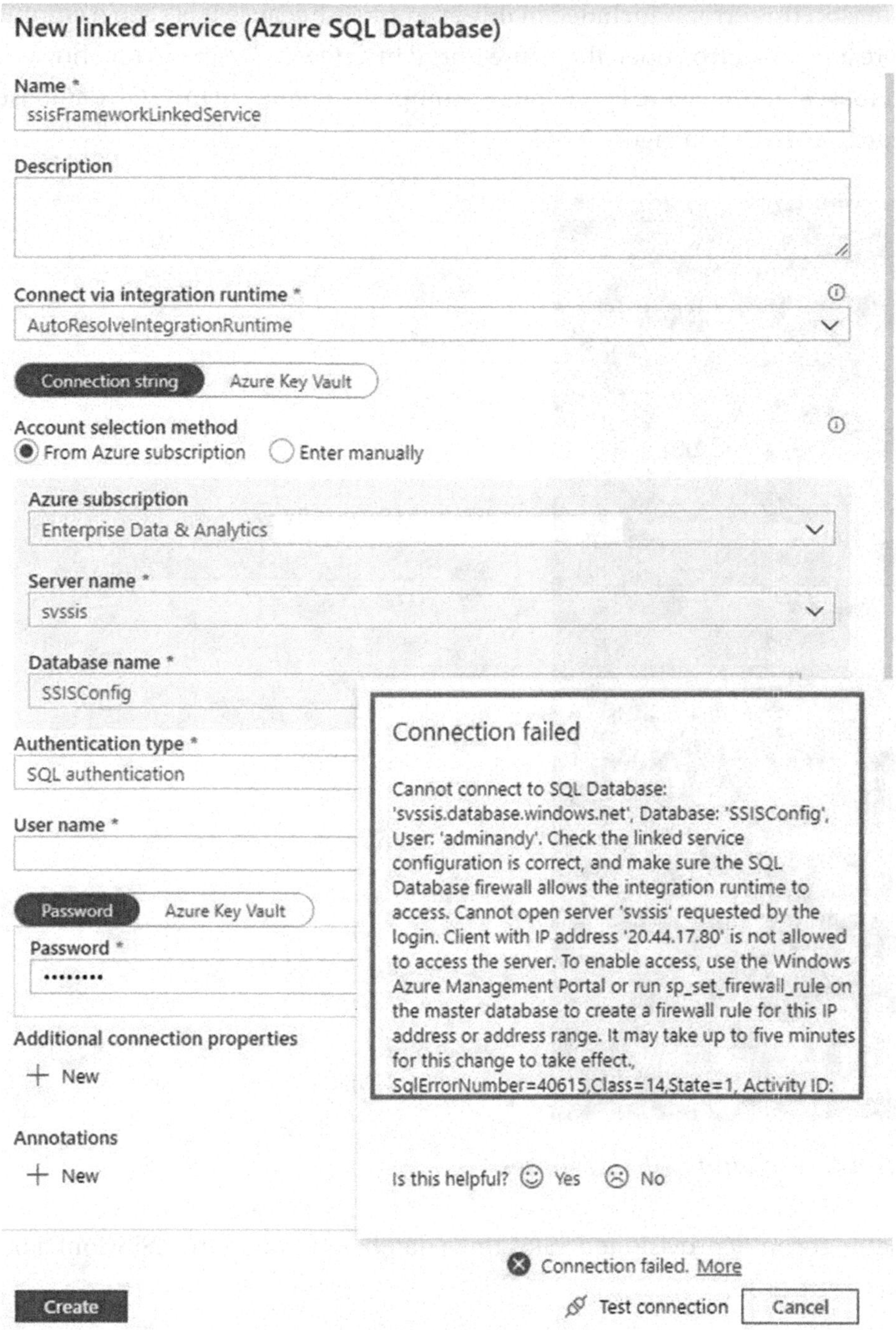

New linked service (Azure SQL Database)

Name *

ssisFrameworkLinkedService

Description

Connect via integration runtime *   ⓘ

AutoResolveIntegrationRuntime   ⌄

(**Connection string**   Azure Key Vault)

Account selection method   ⓘ
◉ From Azure subscription   ◯ Enter manually

Azure subscription

Enterprise Data & Analytics   ⌄

Server name *

svssis   ⌄

Database name *

SSISConfig

Authentication type *

SQL authentication

User name *

(**Password**   Azure Key Vault)

Password *

········

Additional connection properties

➕ New

Annotations

➕ New

Connection failed

Cannot connect to SQL Database:
'svssis.database.windows.net', Database: 'SSISConfig',
User: 'adminandy'. Check the linked service
configuration is correct, and make sure the SQL
Database firewall allows the integration runtime to
access. Cannot open server 'svssis' requested by the
login. Client with IP address '20.44.17.80' is not allowed
to access the server. To enable access, use the Windows
Azure Management Portal or run sp_set_firewall_rule on
the master database to create a firewall rule for this IP
address or address range. It may take up to five minutes
for this change to take effect.,
SqlErrorNumber=40615,Class=14,State=1, Activity ID:

Is this helpful? ☺ Yes   ☹ No

✖ Connection failed. More

∮ Test connection   Cancel

Create

Figure 9-33. *Failed connection test*

The reason this error is included in this example is because it is such a common error. To resolve the error, open the Azure Portal in a new browser tab and hover over "SQL databases" in the Azure left menu to prompt the display of the SQL databases "hover card," as shown in Figure 9-34.

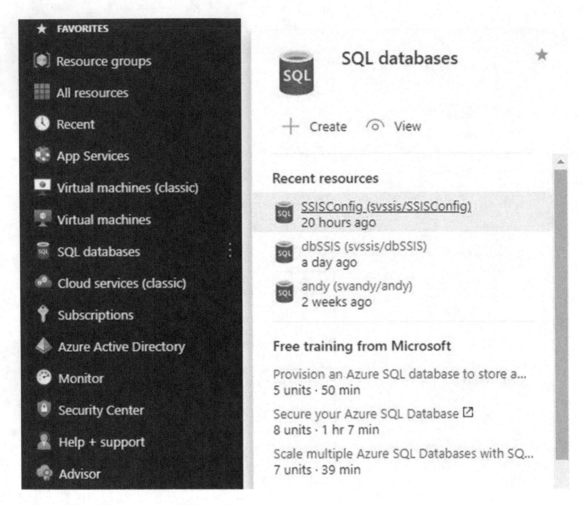

Figure 9-34. *Viewing SQL databases*

Click the recently provisioned SSISConfig database to open the SSISConfig page, as shown in Figure 9-35.

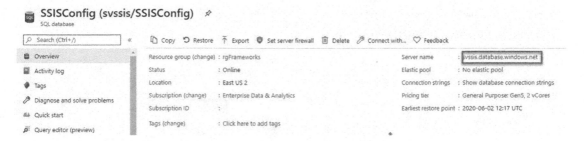

Figure 9-35. *The SSISConfig Azure SQL database blade*

On the Overview page, click the Server name link ("svssis.database.windows.net" in this case) – shown in Figure 9-35 – to open the server blade, as shown in Figure 9-36.

Figure 9-36. *The server that hosts the SSISConfig Azure SQL database*

When the server overview page displays, click the "Show firewall settings" link to open the "Firewalls and virtual networks" page. Change the "Allow Azure services and resources to access this server" to "Yes," as shown in Figure 9-37.

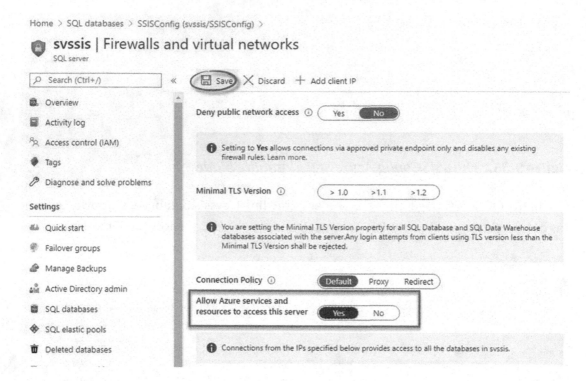

Figure 9-37. *Updating "Allow Azure services and resources to access this server"*

Click the Save button shown in Figure 9-37 to store the update to "Allow Azure services and resources to access this server," and then return to the "New linked service" configuration in Azure Data Factory. Make sure you complete the Linked Service configuration (in the following image, the "User name" property value is redacted). Click "Test connection" to retest the connection to the SSISConfig database, similar to Figure 9-38.

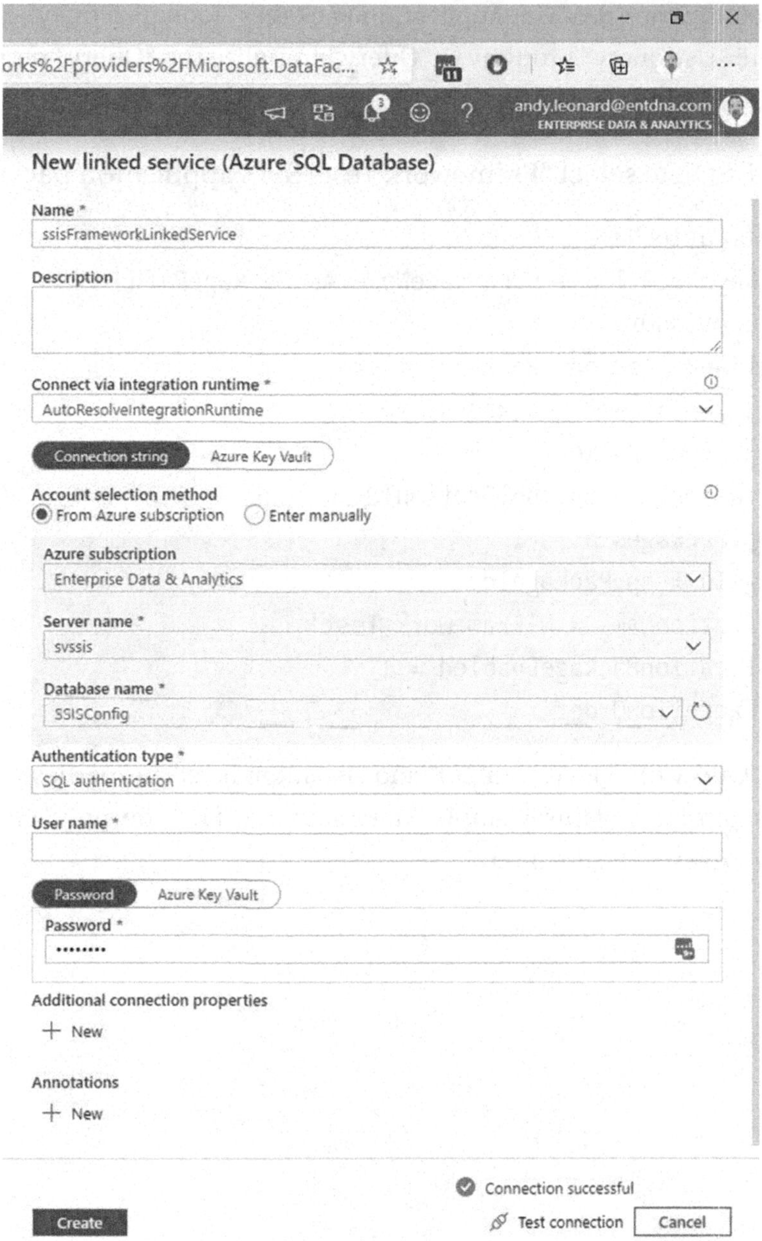

Figure 9-38. *Successful connection test*

The preceding section has not only served as an example of FOA (flexible order of authoring), the section also provided one solution to a very common Azure security obstacle.

Continue configuring the "Get Application Packages" lookup activity's Settings tab by changing the "Use query" property to "Query." Initialize the "Query" property with the T-SQL statement shown in Listing 9-6.

Listing 9-6. T-SQL to select "Framework Test" SSIS application packages

```
Select a.ApplicationName
     , p.PackageLocation + p.PackageName As PackagePath
     , ap.ExecutionOrder
     , ap.FailApplicationOnPackageFailure
From [config].[ApplicationPackages] ap
Join [config].[Applications] a
  On a.ApplicationId = ap.ApplicationId
Join [config].Packages p
  On p.PackageId = ap.PackageId
Where a.ApplicationName = N'Framework Test'
  And ap.ApplicationPackageEnabled = 1
Order By ap.ExecutionOrder
```

Leave the "Query timeout (minutes)" and "Isolation level" properties set to their defaults for the purposes of this example. Make sure the "First row only" checkbox is unchecked, as shown in Figure 9-39.

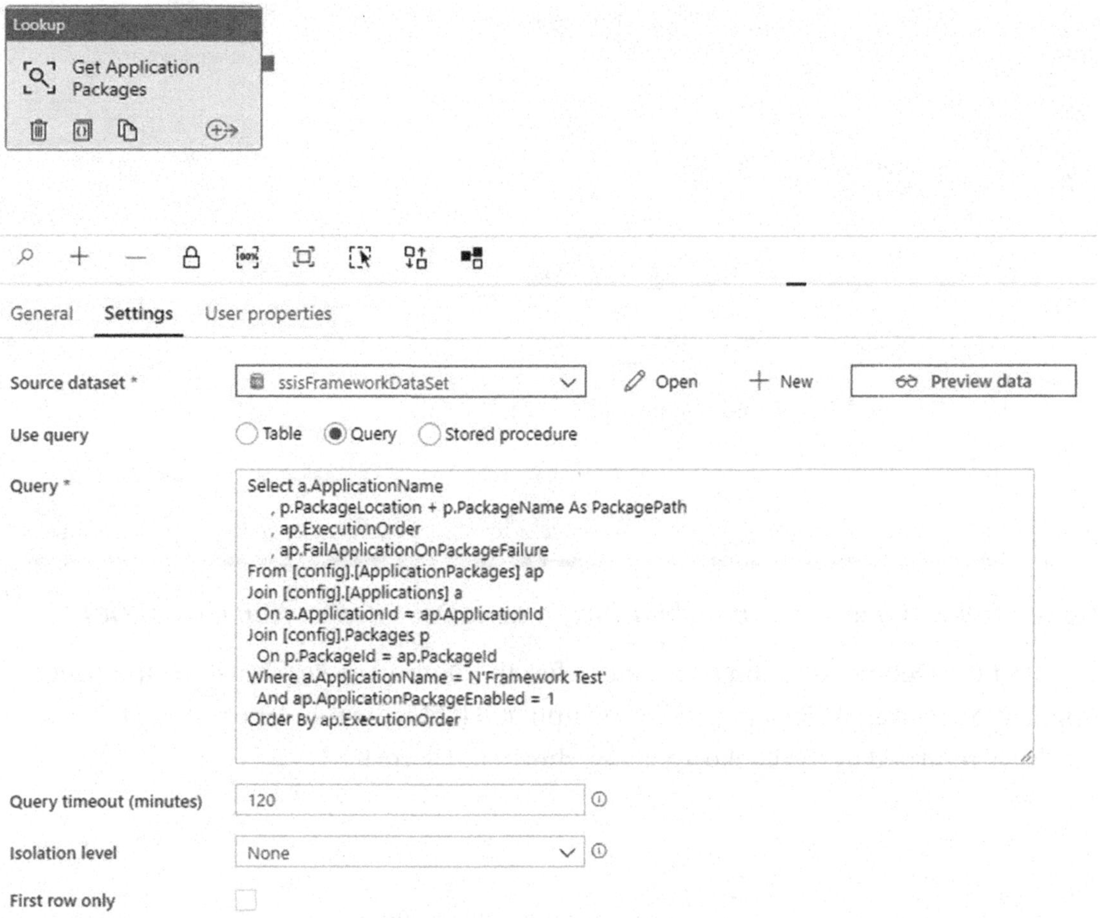

Figure 9-39. *Completing "Get Application Packages" lookup activity settings*

The "Get Application Packages" lookup activity may now be tested by clicking the Debug item in the toolbar, as shown in Figure 9-40.

Figure 9-40. *View the output (blue box) after a Debug execution (red circle)*

After the Debug execution completes, click the Output viewer shown on the "Get Application Packages" lookup activity's Output tab ("blue box" in Figure 9-40) to view the JSON returned by the Lookup activity, shown in Figure 9-41.

```
{
    "count": 2,
    "value": [
        {
            "ApplicationName": "Framework Test",
            "PackagePath": "\\\\stframeworks.file.core.windows.net\\fs-ssis\\ReportAndSucceed.dtsx",
            "ExecutionOrder": 10,
            "FailApplicationOnPackageFailure": true
        },
        {
            "ApplicationName": "Framework Test",
            "PackagePath": "\\\\stframeworks.file.core.windows.net\\fs-ssis\\ReportAndFail.dtsx",
            "ExecutionOrder": 20,
            "FailApplicationOnPackageFailure": true
        }
    ],
    "effectiveIntegrationRuntime": "DefaultIntegrationRuntime (East US 2)",
    "billingReference": {
        "activityType": "PipelineActivity",
        "billableDuration": [
            {
                "meterType": "AzureIR",
                "duration": 0.016666666666666666,
                "unit": "DIUHours"
            }
        ]
    },
    "durationInQueue": {
        "integrationRuntimeQueue": 0
    }
}
```

Figure 9-41. *JSON returned by "Get Application Packages" lookup activity*

Now that we have a list of SSIS packages, the next step is to execute the SSIS packages contained in the list.

Execute the Retrieved SSIS Packages

In the SSIS version of the simple, custom SSIS framework, a Foreach Loop Container enumerates the collection of SSIS packages. In Azure Data Factory, a ForEach activity performs the same function.

Expand the Activities category named "Iteration & conditionals," and drag a ForEach activity onto the pipeline canvas. Connect a Success pipe from the "Get Application Packages" lookup activity to the ForEach activity, and then rename the ForEach activity to "ForEach Application Package," as shown in Figure 9-42.

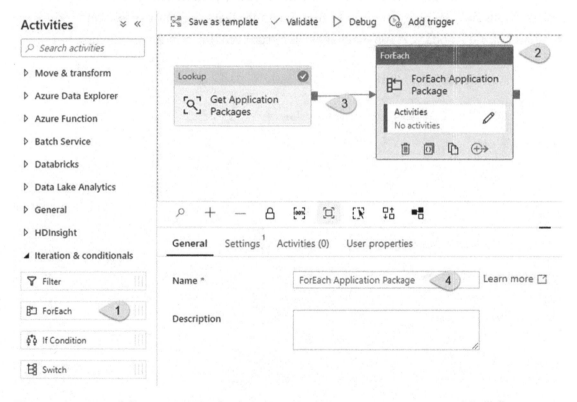

Figure 9-42. *Adding a ForEach activity*

On the Settings tab, click inside the Items property textbox. The "Add dynamic content [Alt + P]" link is displayed just below the Items textbox, as shown in Figure 9-43.

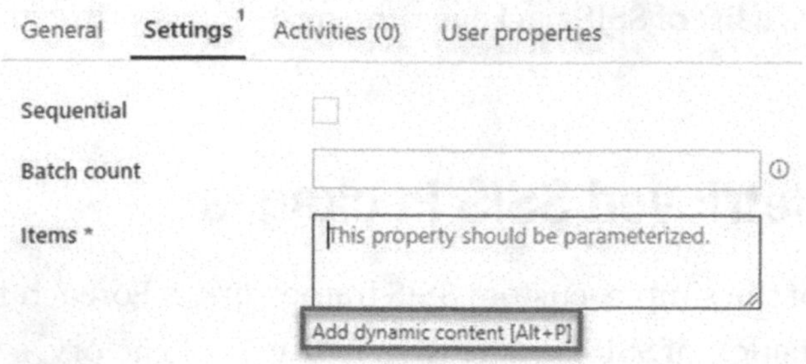

Figure 9-43. *The "Add dynamic content [Alt + P]" link on the "ForEach Application Package" ForEach activity's Settings tab*

Click the "Add dynamic content [Alt + P]" link to open the "Add dynamic content" blade. In the "Activity outputs" category, click Get Application Packages. `@activity('Get Application Packages').output` appears in the Add dynamic content textbox. Append ".value" to the expression so that the Azure Data Factory expression reads `@activity('Get Application Packages').output.value`, as shown in Figure 9-44.

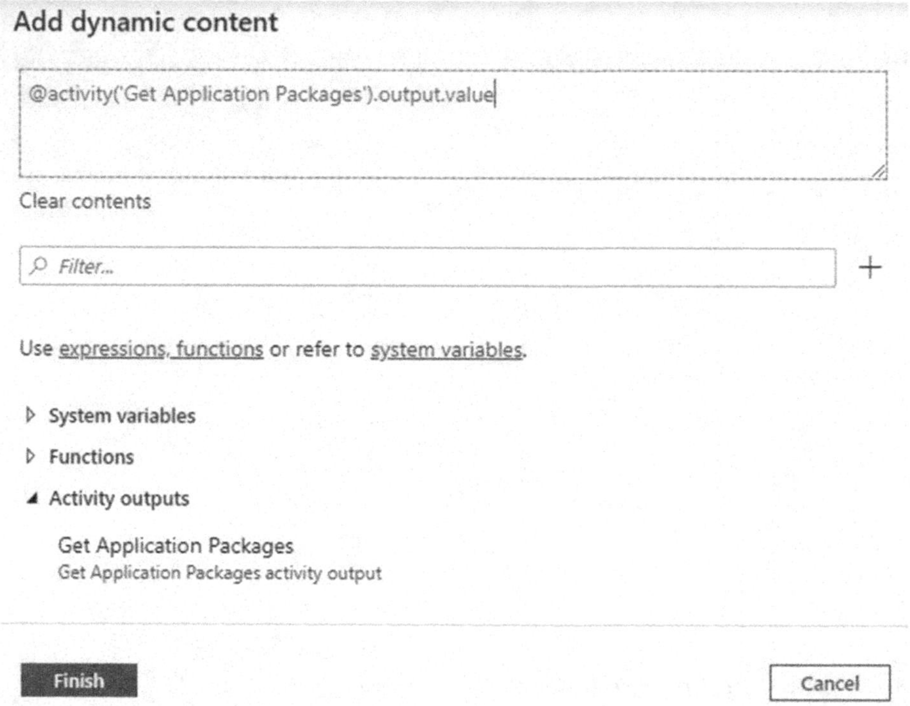

Figure 9-44. *Setting the items property dynamic value for the ForEach Application Packages activity*

Click the Finish button to complete the property configuration. The ForEach Application Packages Items property should now appear similar to Figure 9-45.

General	**Settings**	Activities (0)	User properties

Sequential	☐
Batch count	_____ ⓘ
Items	@activity('Get Application Packages').output.value

Figure 9-45. *The ForEach Application Packages Items property, configured*

The enumerator is now configured. The next step is to configure the activities we wish the parent pipeline to perform for each application package.

Click ForEach Application Package's activities Edit icon (the pencil), shown in Figure 9-46.

Figure 9-46. *Edit the ForEach Application Package's activities*

When the parent ➤ ForEach Application Package ➤ Activities page displays, drag an "Execute SSIS package" activity from the Activities ➤ General pipeline activities, and then rename the new Execute SSIS package activity to "Execute Application Package," as shown in Figure 9-47.

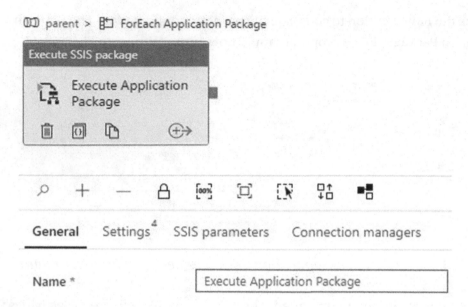

Figure 9-47. *Adding and renaming the "Execute Application Package" execute SSIS package activity*

Configure the "Execute Application Package" execute SSIS activity's Settings tab Azure-SSIS IR property, selecting the "Azure-SSIS-Files" integration runtime from the drop-down, as shown in Figure 9-48.

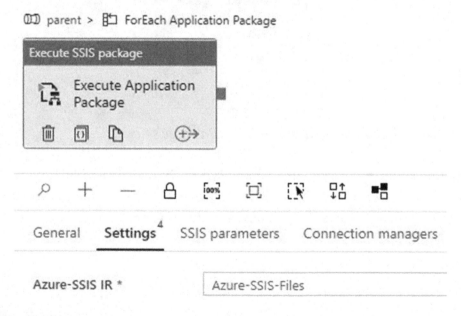

Figure 9-48. *Selecting the Azure-SSIS-Files integration runtime*

For the purposes of this example, ensure the Windows authentication and 32-bit runtime property checkboxes are unchecked. Click inside the "Package path" property's textbox, and then click the "Add dynamic content [Alt + P]" link to surface the Add dynamic content blade, expand the ForEach iterator category, click the ForEach Application Package Current Item option, and then append ".PackagePath" to "@item()" in the dynamic content textbox, as shown in Figure 9-49.

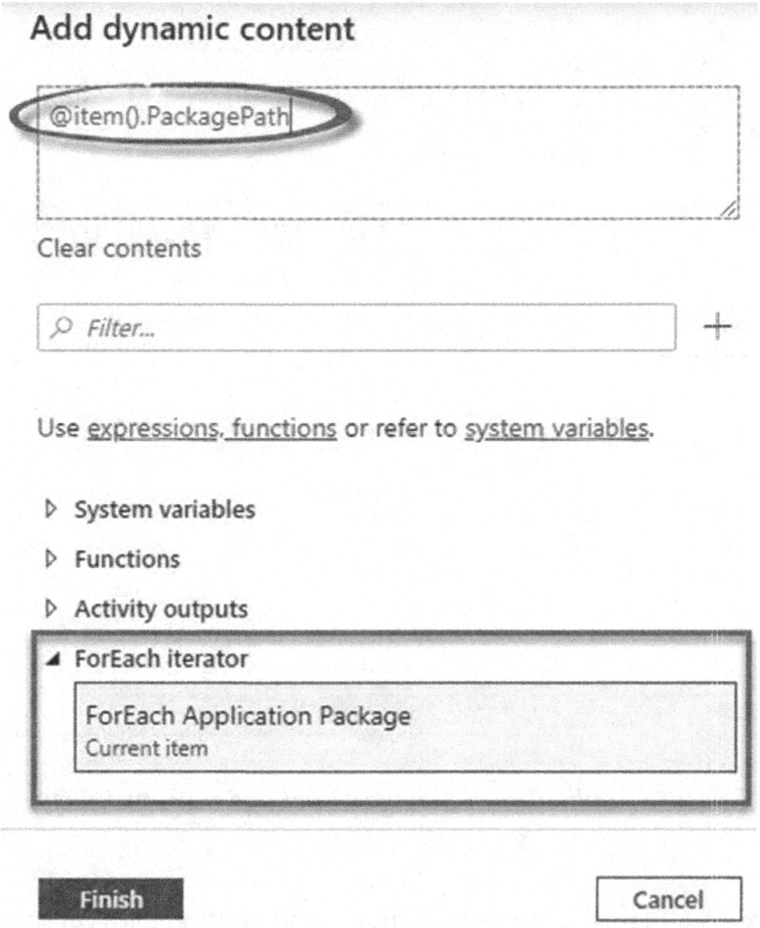

Figure 9-49. *Configuring the PackagePath value in ForEach Application Package activities*

Click the Finish button to return to the Settings tab, as shown in Figure 9-50.

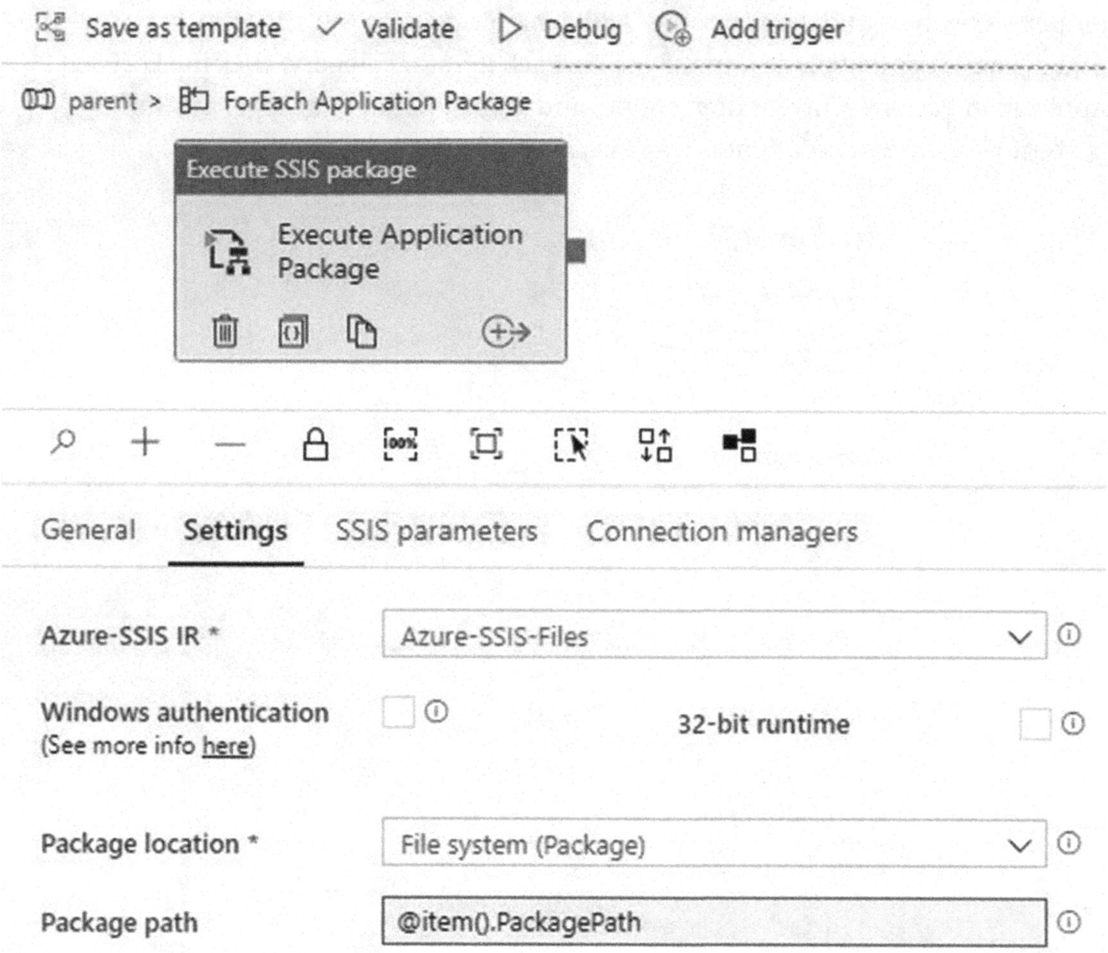

Figure 9-50. Execute Application Package's Package path property, configured

For the purposes of this example, leave the "Configuration path" property empty, although it contains "sample" configuration metadata, set the Domain property to "Azure," and set the Username property to the name of the Azure File Share – "stframeworks" in this case, as shown in Figure 9-51.

Configuration path	\\FileShare\ConfigurationName.dtsConfig	ⓘ

Package access credentials ⓘ
(See more info here)

Domain *	Azure	ⓘ

Username *	stframeworks	ⓘ

Figure 9-51. *Configuring the Execute SSIS package Configuration path, Domain, and Username properties*

Configuring the Password property is the next step, and it is perhaps the most unintuitive property of all. The value required for the Execute SSIS package activity is not a *password,* it is a *key* for accessing the Azure file share. To obtain this value, open Microsoft Azure Storage Explorer and connect to the Azure file share. In the Explorer pane, right-click the Azure file share, and then click "Copy Primary Key," as shown in Figure 9-52.

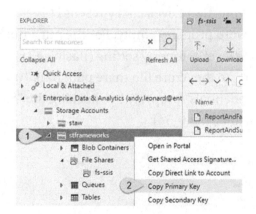

Figure 9-52. *Copying the Azure file share's primary key for access*

The Azure file share primary key is the value required for the password property. Return to the Azure Data Factory portal, and paste the Azure file share primary key into the Password property textbox, as shown in Figure 9-53.

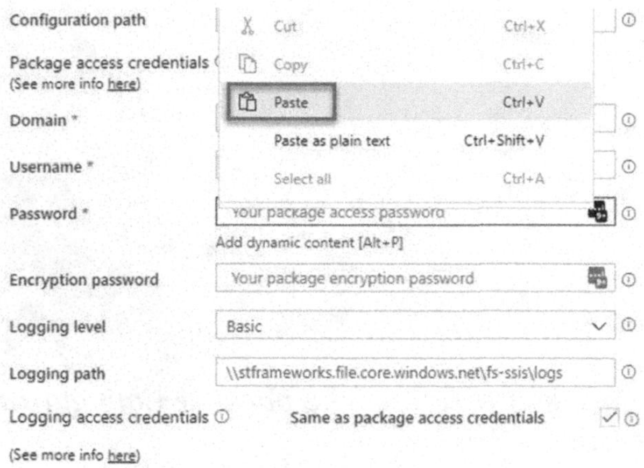

Configuration path			ⓘ
Package access credentials (	✂ Cut	Ctrl+X	
(See more info here)	⬜ Copy	Ctrl+C	
Domain *	📋 Paste	Ctrl+V	ⓘ
Username *	Paste as plain text	Ctrl+Shift+V	ⓘ
	Select all	Ctrl+A	
Password *	Your package access password		🗖 ⓘ

Add dynamic content [Alt+P]

Encryption password	Your package encryption password	🗖 ⓘ
Logging level	Basic	∨ ⓘ
Logging path	\\stframeworks.file.core.windows.net\fs-ssis\logs	ⓘ
Logging access credentials ⓘ	Same as package access credentials	☑ ⓘ

(See more info here)

Figure 9-53. *Pasting the Azure file share primary key into the Execute SSIS package activity's Password property*

If the SSIS package's Package Protection Level property is set to EncryptAllWithPassword or EncryptSensitiveWithPassword, the password in the SSIS package's Package Password property must be supplied to the Execute SSIS package activity's "Encryption password" property (#1 in Figure 9-54). Since our test SSIS project packages use the default Protection Level, EncryptSensitiveWithUserKey, we may leave the Encryption password property empty. For the purposes of this example, accept the default "Logging level" property setting (Basic - #2 in Figure 9-54) and set the "Logging path" property to the Azure file share path plus "\logs" (#3 in Figure 9-54). Set the "Logging access credentials" property by checking the "Same as package access credentials" checkbox (#4 in Figure 9-54).

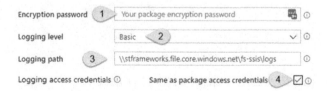

Figure 9-54. *Configuring the Encryption password, Logging level, Logging path, and Logging access credentials properties for the Execute SSIS package activity*

The parent pipeline is now minimally configured, enough so we can click the Debug menu item and perform a test execution (make sure the Azure-SSIS-Files integration runtime is running first!). By design (at this point), the pipeline debug execution should fail, as shown in Figure 9-55.

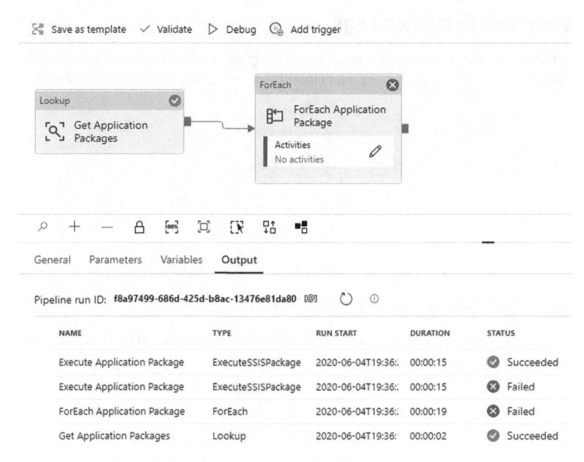

Figure 9-55. Failed test execution or parent pipeline

The test execution failed, but why?

In Figure 9-55, the first execution of the "Execute Application Package" execute SSIS activity failed, but the second execution of the "Execute Application Package" execute SSIS activity succeeded. You may see *both* executions fail. The most likely cause is the Azure-SSIS-Files integration runtime is not running.

The next step is to view the log file generated in the Azure file share's logs folder.

View Test Execution Logs

Returning to storage explorer, the logs folder is likely *not* displayed. Refresh the Azure file share from the "More" drop-down menu item. Click "Refresh," as shown in Figure 9-56.

Figure 9-56. *Clicking More ➤ Refresh to view the logs folder*

The logs folder, shown in Figure 9-57, should have been created (if the logs folder did not exist beforehand) when the test execution occurred.

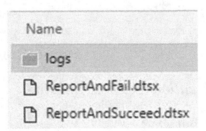

Figure 9-57. *The logs folder, surfaced*

Double-click the logs folder to view the contents. Each SSIS package execution generates a run ID visible in the pipeline's Output tab in ADF, as shown in Figure 9-58, along with the corresponding logs\folder name in storage explorer.

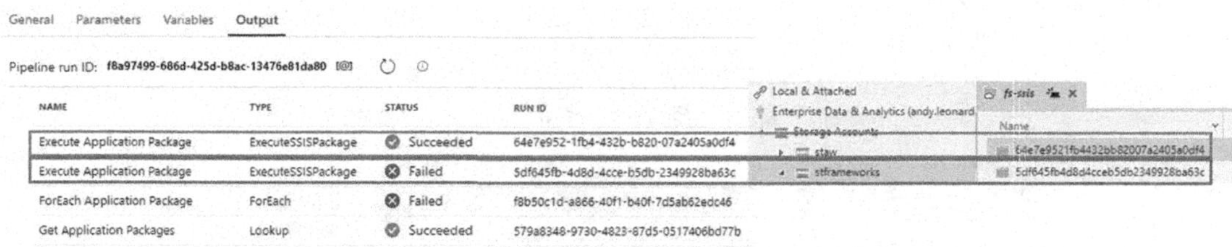

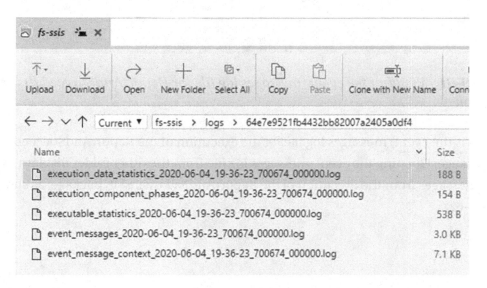

Figure 9-58. *Pipeline execution's Output tab aligned with Storage Explorer's logs folder contents*

"Execute Application Package" executed twice, once to run the SSIS package named "ReportAndSucceed.dtsx" and again for the SSIS package named "ReportAndFail.dtsx". Figure 9-55 showed the topmost instance of "Execute Application Package" succeeded, and the next listing of "Execute Application Package" failed.

Figure 9-58 is a screenshot of the parent pipeline's Output tab *aligned* with the contents of the logs folder in the Azure file share. Please note the pipeline's Output Run ID value is similar to the name of the logs subfolder. Entering the topmost logs subfolder in storage explorer reveals a collection of log files, as shown in Figure 9-59.

Figure 9-59. *Log files stored in the Azure file share fs-ssis\logs folder*

Right-click the file that begins with "event_messages_" and then click "Open," as shown in Figure 9-60.

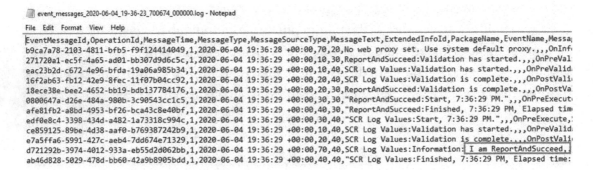

Figure 9-60. *Opening the event_messages log*

Since my server is configured to open files with the ".log" extension using Notepad, the event_messages file opens in Notepad, as shown in Figure 9-61.

Figure 9-61. *Viewing the event_messages log file for ReportAndSucceed.dtsx test execution*

This is the event_messages log file for the execution of the ReportAndSucceed.dtsx SSIS package. Opening the event_messages log file from the other folder displays the event_messages from the execution of the ReportAndFail.dtsx SSIS package, as shown in Figure 9-62.

```
event_messages_2020-06-04_19-36-23_700674_000000.log - Notepad

File  Edit  Format  View  Help

EventMessageId,OperationId,MessageTime,MessageType,MessageSourceType,MessageText,ExtendedInfoId,PackageName,EventName,MessageSor
132eefc8-de7d-4a6a-9e2f-8e862e37159b,1,2020-06-04 19:36:28 +00:00,70,20,No web proxy set. Use system default proxy.,,,OnInforma
214cf0ba-6ffb-4755-922e-ade15154fd16,1,2020-06-04 19:36:29 +00:00,10,30,ReportAndFail:Validation has started.,,,OnPreValidate,R
b1430748-a515-470e-9093-1c40521a8add,1,2020-06-04 19:36:29 +00:00,10,40,SCR Log Values:Validation has started.,,,OnPreValidate,
335bb3be-2782-45fd-ae7d-1313c4bf382e,1,2020-06-04 19:36:29 +00:00,20,40,SCR Log Values:Validation is complete.,,,OnPostValidate
6dc5508e-d3e0-4325-9e4f-8f058cb3d0e6,1,2020-06-04 19:36:29 +00:00,20,30,ReportAndFail:Validation is complete.,,,OnPostValidate,
78fe5a00-ad5a-4425-bf1c-bbd72e46195b,1,2020-06-04 19:36:29 +00:00,30,30,"ReportAndFail:Start, 7:36:29 PM.",,,OnPreExecute,Repor
8d61c66d-356e-44d4-b052-933cff096c11,1,2020-06-04 19:36:29 +00:00,110,30,"ReportAndFail:Warning: SSIS Warning Code DTS_W_MAXIMUI
695f060b-9416-422c-a477-b01edadee1c2,1,2020-06-04 19:36:29 +00:00,40,30,"ReportAndFail:Finished, 7:36:29 PM, Elapsed time: 00:0
d8c121a3-fd33-4f55-8c99-aa337137f32c,1,2020-06-04 19:36:29 +00:00,30,40,"SCR Log Values:Start, 7:36:29 PM.",,,OnPreExecute,SCR
d78979ef-37ab-4839-81af-8beb00d5e9e9,1,2020-06-04 19:36:29 +00:00,10,40,SCR Log Values:Validation has started.,,,OnPreValidate,
6b45c231-3428-43b3-af99-3593497e0e55,1,2020-06-04 19:36:29 +00:00,20,40,SCR Log Values:Validation is complete.,,,OnPostValidate
e4fc00eb-fd2d-4019-9275-51375590cf07,1,2020-06-04 19:36:29 +00:00,120,40,SCR Log Values:Error: ReportAndFail execution failed,,
d1296801-f67c-4833-9aa3-0677bc61e80d,1,2020-06-04 19:36:29 +00:00,110,40,"SCR Log Values:Warning: SSIS Warning Code DTS_W_MAXIM
93e42c01-fb54-4e2f-87c5-7d17b7a054bc,1,2020-06-04 19:36:29 +00:00,130,40,"SCR Log Values:The task, 'SCR Log Values', failed.",,
a4f07233-f9c1-4221-80c8-d703dacfc943,1,2020-06-04 19:36:29 +00:00,40,40,"SCR Log Values:Finished, 7:36:29 PM, Elapsed time: 00:
```

Figure 9-62. *Viewing the event_messages log file for ReportAndFail.dtsx test execution*

Conclusion

This chapter covered the following:

- Provisioning an Azure SQL Database instance

  - Deploying the SSISConfig database to the Azure SQL DB instance

- Provisioning an Azure File Share

  - Deploying (copying) test SSIS packages to the Azure file share

- Building the SSIS framework execution engine

  - Test-executing the parent pipeline and viewing execution logs

The next step is persisting application and application package operational data in the SSISConfig database.

CHAPTER 10

Framework Logging in ADF

Chapters 5–7 covered designing and building an SSIS framework using a SQL Server database named SSISConfig and an SSIS package as the execution engine (the Parent. dtsx SSIS package) to execute SSIS packages stored in the file system of an on-premises server. Chapter 8 covered provisioning an instance of Azure Data Factory (ADF), an Azure storage account, and an Azure-SSIS integration runtime configured to execute SSIS packages stored in an Azure file share. Chapter 9 covered provisioning an Azure SQL database, deploying the SSISConfig database to the Azure SQL DB, provisioning an Azure file share (fs-ssis), and then building the execution engine (the parent pipeline) in ADF.

In this chapter, we add logging functionality to the parent ADF pipeline, much like the functionality we added to the on-premises version of the SSIS framework in Chapter 7. The log schema and associated artifacts have already been deployed, but we will make a few adjustments and modifications to the logging-related tables.

Although they perform similar operations, Azure Data Factory pipelines are *different* from SSIS packages. The differences will drive changes in the SSIS framework design.

Add the ApplicationName Parameter

The parent pipeline will execute SSIS framework applications. To make the ADF pipeline generic and able to execute *any* application stored in the SSISConfig database's metadata, the pipeline requires an ApplicationName parameter.

Open the parent pipeline, click the whitespace of the pipeline, and then click the Parameters tab. When the Parameters tab displays, click the "+ New" link highlighted in Figure 10-1.

301

A. Leonard and K. Bradshaw, *SQL Server Data Automation Through Frameworks*,
https://doi.org/10.1007/978-1-4842-6213-9_10

Figure 10-1. *Beginning to add a new parameter*

When the Parameter configuration displays, set the following parameter properties, which are shown in Figure 10-2.

- Name: ApplicationName

- Type: String

- Default Value: Framework Test

Figure 10-2. *Configuring the ApplicationName parent pipeline parameter*

The ApplicationName parameter is now accessible by activities in the parent ADF pipeline.

A Quick Review of SSIS Framework Applications and Packages

An SSIS framework application is a collection of SSIS packages that execute in a prescribed order (*one* SSIS application may be configured to execute *one or many* SSIS packages). Metadata for the SSIS application is stored in the config.Applications table and metadata for SSIS package location is stored in the config.Packages table. Because the cardinality between SSIS applications and SSIS packages is many-to-many (*one* SSIS package may be configured to run in *one or many* SSIS applications), the config.ApplicationPackages table exists to resolve relationships between applications and packages.

Add Application Instance Logging

In the introduction to this chapter, I mentioned there are differences between the way ADF and SSIS accomplish similar functions, and these differences will drive changes in the SSIS framework design in Azure Data Factory. We encounter the first difference here as we configure Application Instance logging.

The SSIS framework collects operational data about instances of application execution in the log.ApplicationInstance table. Operational data regarding instances of application package execution are stored in the log.ApplicationPackageInstance table.

Modifying the Log.ApplicationInstance Table

An Azure Data Factory (ADF) pipeline is made up of many constituent objects and artifacts, including ADF activities. When ADF activities execute, a *RunId* value is generated for use in operational logging. When an Azure Data Factory pipeline executes, a RunId value is generated and used to connect internal ADF pipeline logs to pipeline activity logs.

Modify the log.ApplicationInstance table by executing the T-SQL in Listing 10-1.

Listing 10-1. Adding the ApplicationRunId column to the ApplicationInstance table

```
print 'Log.ApplicationInstance.ApplicationRunId column'
If Not Exists(Select s.[name]
            + '.' + t.[name]
            + '.' + c.[name]
          From [sys].[schemas] s
          Join [sys].[tables] t
            On t.[schema_id] = s.[schema_id]
          Join [sys].[columns] c
            On c.[object_id] = t.[object_id]
          Where s.[name] = N'log'
            And t.[name] = N'ApplicationInstance'
            And c.[name] = N'ApplicationRunId')
  begin
   print ' - Adding log.ApplicationInstance.ApplicationRunId column'
   Alter Table log.ApplicationInstance
```

```
      Add ApplicationRunId nvarchar(55) NULL
    print ' - Log.ApplicationInstance.ApplicationRunId column added'
    end
Else
  begin
    print ' - Log.ApplicationInstance.ApplicationRunId column already
    exists.'
  end
```

Let's next encapsulate the T-SQL for adding Application Instance rows in a stored procedure.

Adding the Log.InsertApplicationInstance Stored Procedure

In the SSIS execution engine (the Parent.dtsx SSIS package), an Execute SQL Task was used to insert the initial ApplicationInstance record into the log.ApplicationInstance table and return an ApplicationInstanceId value. In the Azure Data Factory execution engine (the parent pipeline), a Lookup activity will call a stored procedure that inserts the record and returns an ApplicationInstanceId value. The ApplicationInstanceId value will be used later in the pipeline to update the ApplicationInstance record.

In Azure Data Studio (or SSMS), connect to the Azure SQL database named SSISConfig. Starting with the T-SQL code from Listing 7-4 in Chapter 7, execute the T-SQL to create a new stored procedure named "log.InsertApplicationInstance" in an idempotent (re-executable) manner, as shown in Listing 10-2.

Listing 10-2. Idempotent T-SQL to create the log.InsertApplicationInstance stored procedure

```
print 'log.InsertApplicationInstance stored procedure'
If Exists(Select s.[name] + '.' + p.[name]
          From [sys].[procedures] p
          Join [sys].[schemas] s
            On s.[schema_id] = p.[schema_id]
        Where s.[name] = N'log'
          And p.[name] = N'InsertApplicationInstance')
```

```
begin
  print ' - Dropping log.InsertApplicationInstance stored procedure'
  Drop Procedure log.InsertApplicationInstance
  print ' - Log.InsertApplicationInstance stored procedure dropped'
end

print ' - Creating log.InsertApplicationInstance stored procedure'
go

Create Procedure log.InsertApplicationInstance
   @ApplicationName nvarchar(255)
 , @ApplicationRunId nvarchar(55) = NULL
As

  declare @ApplicationId int = (Select ApplicationId
                                 From config.Applications
                                 Where ApplicationName = @ApplicationName)

  Insert Into [log].ApplicationInstance (ApplicationId, ApplicationRunId)
  Output inserted.ApplicationInstanceId
  Values (@ApplicationId, @ApplicationRunId)

go

print ' - Log.InsertApplicationInstance stored procedure created'
go
```

The log.InsertApplicationInstance stored procedure requires the name of an SSIS framework application be sent to the @ApplicationName string [nvarchar(255)] parameter. An internal T-SQL int parameter named @ApplicationId looks up the value of the ApplicationId in the config.Applications table for the given @ApplicationName value.

The log.InsertApplicationInstance stored procedure also accepts a parameter named @ApplicationRunId which defaults to NULL when no value is supplied. @ApplicationRunId is used to store the Azure Data Factory pipeline RunId value. Storing the RunId in the SSIS framework allows enterprise DevOps teams to relate operational log data in the SSIS framework to operational data captured in ADF logs, which is *extremely* useful when something unfortunate happens during ADF pipeline execution.

The T-SQL `Insert` statement initializes the ApplicationInstance record by inserting new `@ApplicationId` and `@ApplicationRunId` values into the log.ApplicationInstance table. Defaults configured on the log.ApplicationInstance table insert the following values in the row:

- `ApplicationStartTime` is set to `sysdatetimeoffset()` by `DF_log_ApplicationInstance_ApplicationStartTime`.

- `ApplicationStatus` is set to `N'Running'` by `DF_log_ApplicationInstance_ApplicationStatus`.

If all goes as planned, Azure Data Studio should return Messages that appear as shown in Figure 10-3.

Messages

```
Started executing query at Line 1
log.InsertApplicationInstance stored procedure
 - Creating log.InsertApplicationInstance stored procedure
Started executing query at Line 16
Commands completed successfully.
Started executing query at Line 31
 - Log.InsertApplicationInstance stored procedure created
Total execution time: 00:00:00.110
```

Figure 10-3. *Messages from creating the log.InsertApplicationInstance stored procedure*

After the log.InsertApplicationInstance stored procedure has been created, the next step is to add an activity to the pipeline to execute the stored procedure.

Logging Application Instance

The ADF execution engine needs to execute a stored procedure, log. InsertApplicationInstance, that returns a value named ApplicationInstanceId. A Lookup activity is built for this!

Return to the parent ADF pipeline. Drag a new Lookup activity onto the surface. Add a Success output from the new Lookup activity to the Get Application Packages lookup activity. Rename the new Lookup activity to "Log Application Instance Start," as shown in Figure 10-4.

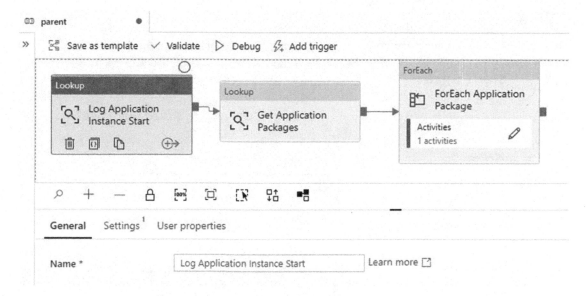

Figure 10-4. *Adding the Log Application Instance Start lookup activity*

Click the Log Application Instance Start lookup activity's Settings tab to continue configuration. Click the "Source dataset" property drop-down, and select "ssisFrameworkDataSet," as shown in Figure 10-5.

Figure 10-5. *Selecting "ssisFrameworkDataSet" for the Log Application Instance Start lookup activity's "Source dataset" property*

Next, set the Log Application Instance Start lookup activity's "Use query" property option to "Stored procedure." Select "log.InsertApplicationInstance" from the "Name" property drop-down, as shown in Figure 10-6.

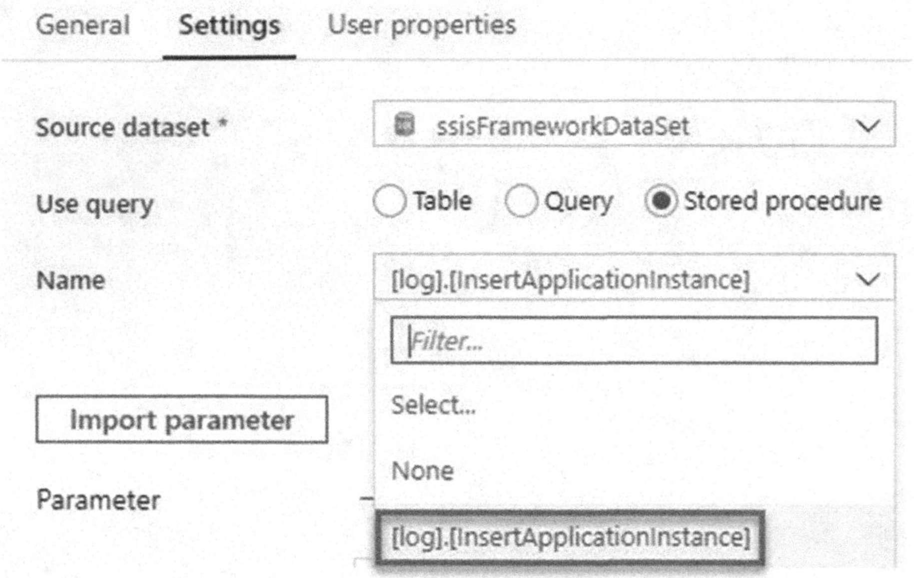

General   **Settings**   User properties

Source dataset *   ssisFrameworkDataSet   ⌄

Use query   ○ Table   ○ Query   ● Stored procedure

Name   [log].[InsertApplicationInstance]   ⌄

        |Filter...

        Select...

Import parameter   None

Parameter   [log].[InsertApplicationInstance]

Figure 10-6. *Selecting the log.InsertApplicationInstance stored procedure*

Next, click the "Import parameter" button to import the @ApplicationName and
@ApplicationRunId String parameters from the log.InsertApplicationInstance stored
procedure. Click inside the @ApplicationName parameter's Value property textbox, and
then click the "Add dynamic content [Alt + P]" link, as shown in Figure 10-7.

Figure 10-7. *Preparing to add dynamic content to the parameter value
property*

When the "Add dynamic content" blade displays, expand the Parameters expression
category, and click "ApplicationName" to set the parameter value ADF expression to
@pipeline.parameters.ApplicationName, as shown in Figure 10-8.

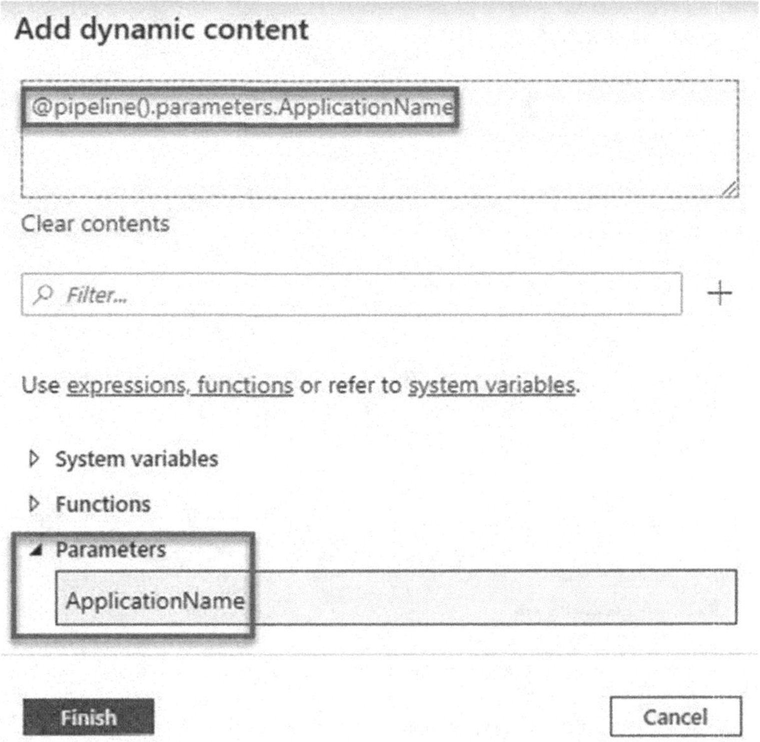

Add dynamic content

@pipeline().parameters.ApplicationName

Clear contents

🔎 Filter...   +

Use expressions, functions or refer to system variables.

▷ System variables

▷ Functions

◢ Parameters

ApplicationName

Finish   Cancel

Figure 10-8. *Setting the ApplicationName parameter value expression*

Click the Finish button to proceed.

Click inside the @ApplicationRunId parameter's Value property textbox, and then click the "Add dynamic content [Alt + P]" link. When the "Add dynamic content" blade displays, expand the "System variables" expression category, and click "Pipeline run ID" to set the parameter value ADF expression to @pipeline().RunId, as shown in Figure 10-9.

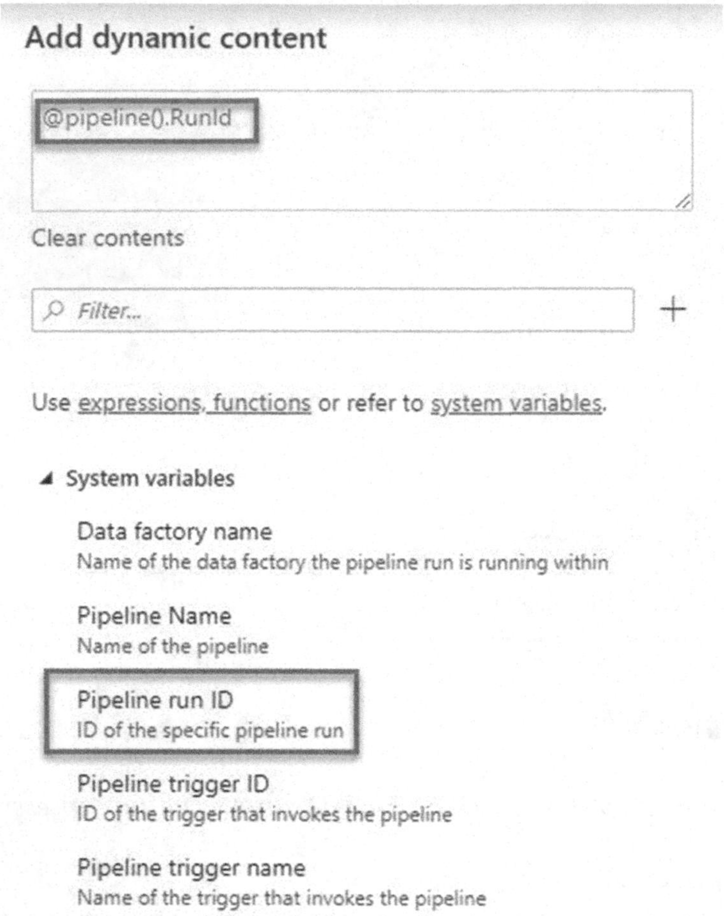

Add dynamic content

@pipeline().RunId

Clear contents

🔍 Filter...   +

Use expressions, functions or refer to system variables.

◢ System variables

    Data factory name
    Name of the data factory the pipeline run is running within

    Pipeline Name
    Name of the pipeline

    Pipeline run ID
    ID of the specific pipeline run

    Pipeline trigger ID
    ID of the trigger that invokes the pipeline

    Pipeline trigger name
    Name of the trigger that invokes the pipeline

Figure 10-9. *Setting the ApplicationRunId parameter value expression*

Leave the "Query timeout (minutes)" property set to its default (120 minutes), the "Isolation level" property set to "None" (default), and the "First row only" property checked (default), as shown in Figure 10-10.

Query timeout (minutes) | 120 | ⓘ

Isolation level | None ∨ | ⓘ

First row only   ☑

Figure 10-10. *Leaving default values for Query timeout, Isolation level, and First row only properties*

The Log Application Instance Start lookup activity is now configured to call the log.InsertApplicationInstance stored procedure, passing the stored procedure the pipeline's ApplicationName and ApplicationRunId parameter values and returning the ApplicationInstanceId value.

Adding the Log.UpdateApplicationInstanceStatus Stored Procedure

As configured, the parent pipeline fails execution because the ReportAndFail.dtsx SSIS package fails (by design). The log.ApplicationInstance row should be updated to reflect the SSIS framework application instance failed.

Begin by adding a new stored procedure named "log. UpdateApplicationInstanceStatus" to the SSISConfig database, as shown in Listing 10-3.

Listing 10-3. Idempotent T-SQL to create the log.UpdateApplicationInstanceStatus stored procedure

```
print 'log.UpdateApplicationInstanceStatus stored procedure'
If Exists(Select s.[name] + '.' + p.[name]
          From [sys].[procedures] p
          Join [sys].[schemas] s
            On s.[schema_id] = p.[schema_id]
          Where s.[name] = N'log'
            And p.[name] = N'UpdateApplicationInstanceStatus')
 begin
  print ' - Dropping log.UpdateApplicationInstanceStatus stored procedure'
  Drop Procedure log.UpdateApplicationInstanceStatus
  print ' - Log.UpdateApplicationInstanceStatus stored procedure dropped'
 end

print ' - Creating log.UpdateApplicationInstanceStatus stored procedure'
go

Create Procedure log.UpdateApplicationInstanceStatus
   @ApplicationInstanceId int
 , @ApplicationStatus nvarchar(55) = 'Succeeded'
As
```

```
Update [log].ApplicationInstance
Set ApplicationEndTime = sysdatetimeoffset()
  , ApplicationStatus = @ApplicationStatus
Where ApplicationInstanceId = @ApplicationInstanceId

go

print ' - Log.UpdateApplicationInstanceStatus stored procedure created'
go
```

If all goes as planned, Azure Data Studio messages should appear similar to Figure 10-11.

Messages

```
Started executing query at Line 1
log.UpdateApplicationInstanceStatus stored procedure
 - Creating log.UpdateApplicationInstanceStatus stored procedure
Started executing query at Line 17
Commands completed successfully.
Started executing query at Line 29
 - Log.UpdateApplicationInstanceStatus stored procedure created
Total execution time: 00:00:00.657
```

Figure 10-11. *Azure Data Studio messages reflecting successful creation of the log. UpdateApplicationInstanceStatus stored procedure*

The next step is to add a Stored Procedure Activity that calls the log.UpdateApplicationInstanceStatus stored procedure on failure.

Updating Application Instance

The ADF execution engine needs to execute a stored procedure, log. UpdateApplicationInstanceStatus that accepts `ApplicationInstanceId` and `ApplicationStatus` values and then updates the status of the current Application Instance. A Stored Procedure Activity is perfect for this!

Return to the parent ADF pipeline. Drag a Stored Procedure Activity onto the surface. Add a Failure output from the ForEach Application Package activity to the new Stored Procedure Activity. Rename the new Stored Procedure Activity to "Log Application Instance Failure," as shown in Figure 10-12.

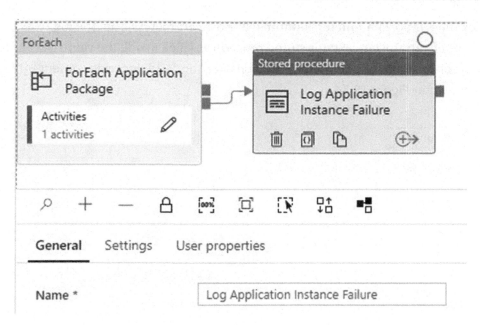

Figure 10-12. Adding the "Log Application Instance Failure" stored procedure activity

On the Settings tab, select "ssisFrameworkLinkedService" from the "Linked service" property drop-down, and then select "[log].[UpdateApplicationInstanceStatus]" from the "Stored procedure name" drop-down, as shown in Figure 10-13.

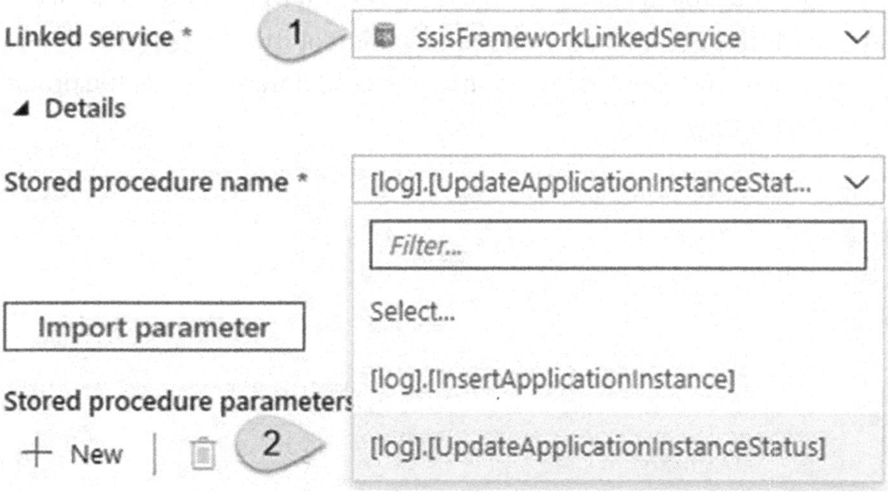

Figure 10-13. Configuring the "Linked service" and "Stored procedure name" properties

Click the "Import parameter" button to import parameters from the log. UpdateApplicationInstanceStatus stored procedure. Click inside the value textbox for the ApplicationInstanceId parameter, and then click the "Add dynamic content [Alt + P]" link, as shown in Figure 10-14.

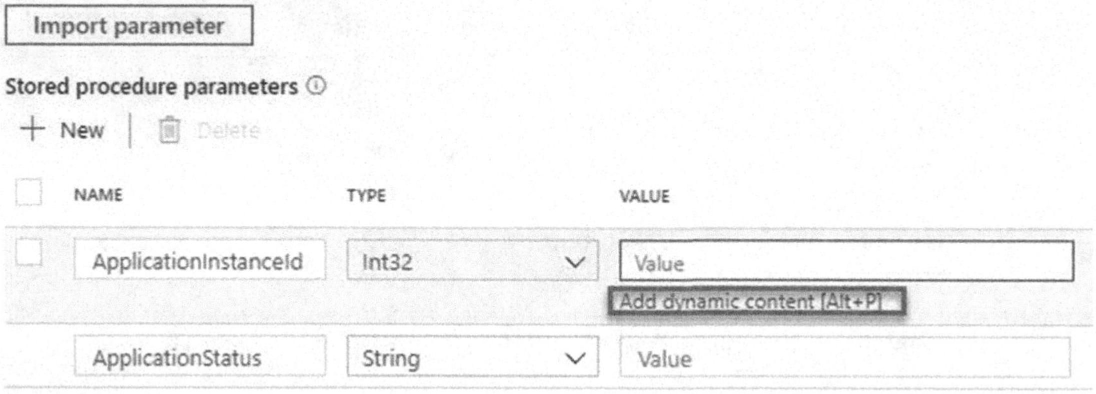

Figure 10-14. *Preparing to enter dynamic content for the ApplicationInstanceId parameter*

When the "Add dynamic content" blade displays, expand the "Activity outputs" expressions category, and click "Log Application Instance Start." The dynamic content textbox will display @activity('Log Application Instance Start').output. Append ".firstrow.ApplicationInstanceId" to the expression to map the value of the ApplicationInstanceId returned from the execution of the "Log Application Instance Start" lookup activity into the ApplicationInstanceId parameter's value property, as shown in Figure 10-15.

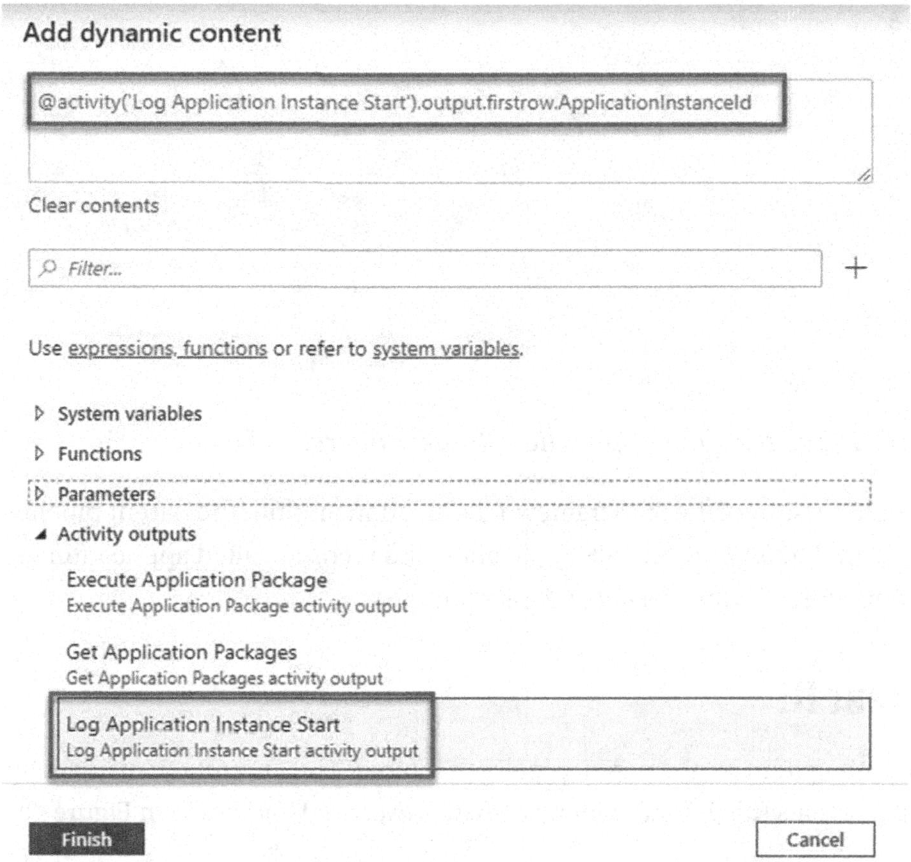

Figure 10-15. *Mapping the "Log Application Instance Start" lookup activity into the ApplicationInstanceId parameter's value property*

Click the Finish button to proceed.

Type "Failed" into the ApplicationStatus parameter's value property, as shown in Figure 10-16.

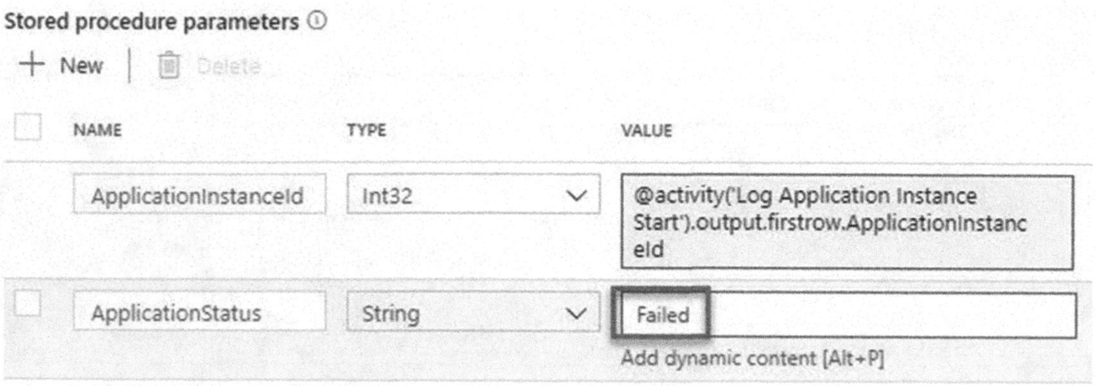

Stored procedure parameters ⓘ

+ New | 🗑 Delete

	NAME	TYPE		VALUE
☐	ApplicationInstanceId	Int32	⌄	@activity('Log Application Instance Start').output.firstrow.ApplicationInstanc eId
☐	ApplicationStatus	String	⌄	Failed

Add dynamic content [Alt+P]

Figure 10-16. *Setting the ApplicationStatus property to Failed*

The ADF version of the SSIS framework execution engine (the parent pipeline) is now configured to execute an SSIS application and record a failed application instance.

The next step is to test-execute the pipeline.

Let's Test It!

Click the Debug item on the parent pipeline's toolbar. The Output tab displays and, after some time, reveals the parent pipeline activities executed, as shown in Figure 10-17.

General   Parameters   Variables   **Output**

NAME	TYPE	DURATION	STATUS
Log Application Instance Failure	SqlServerStoredProcedure	00:00:13	✅ Succeeded
Execute Application Package	ExecuteSSISPackage	00:00:22	✅ Succeeded
Execute Application Package	ExecuteSSISPackage	00:00:26	❌ Failed
ForEach Application Package	ForEach	00:00:32	❌ Failed
Get Application Packages	Lookup	00:00:13	✅ Succeeded
Log Application Instance Start	Lookup	00:00:03	✅ Succeeded

Figure 10-17. *Results of a text execution of the parent pipeline*

Execute the T-SQL query shown in Listing 10-4 to examine the Application Instance results.

Listing 10-4. Application Instance results

```
Select a.ApplicationName
    , ai.ApplicationStartTime
    , ai.ApplicationStatus
From log.ApplicationInstance ai
Join config.Applications a
  On a.ApplicationId = ai.ApplicationId
Order By ApplicationInstanceId Desc
```

The results should appear similar to Figure 10-18.

Results   . Messages

	ApplicationName	ApplicationStartTime	ApplicationStatus
1	Framework Test	2020-06-08 16:22:53…	Failed

Figure 10-18. *Application Instance results*

While the results match what was expected, something is missing: there is no Stored Procedure Activity that updates the Application Instance on Success.

Updating Application Instance on Success

The parent pipeline needs a way to log Application Instance success. Begin by clicking the Clone icon on the "Log Application Instance Failure" stored procedure activity as shown in Figure 10-19.

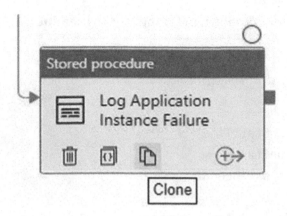

Figure 10-19. *Preparing to clone the "Log Application Instance Success" stored procedure activity from the "Log Application Instance Failure" stored procedure activity*

Cloning the "Log Application Instance Failure" stored procedure activity makes sense because the new Stored Procedure Activity will be calling the *same* stored procedure – log.UpdateApplicationInstanceStatus – and *only one* parameter value configuration will change.

Connect a Success output from the "ForEach Application Package" ForEach activity to the cloned Stored Procedure Activity, and then rename the cloned Stored Procedure Activity to "Log Application Instance Success," as shown in Figure 10-20.

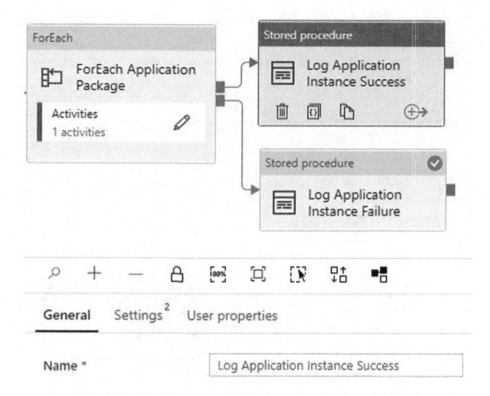

Figure 10-20. *Adding the "Log Application Instance Success" stored procedure activity*

Next, click the Settings tab for the "Log Application Instance Success" stored procedure activity. The only change required on the Settings tab is the value of the ApplicationStatus parameter's value property, which needs to be updated from "Failed" to "Succeeded," as shown in Figure 10-21.

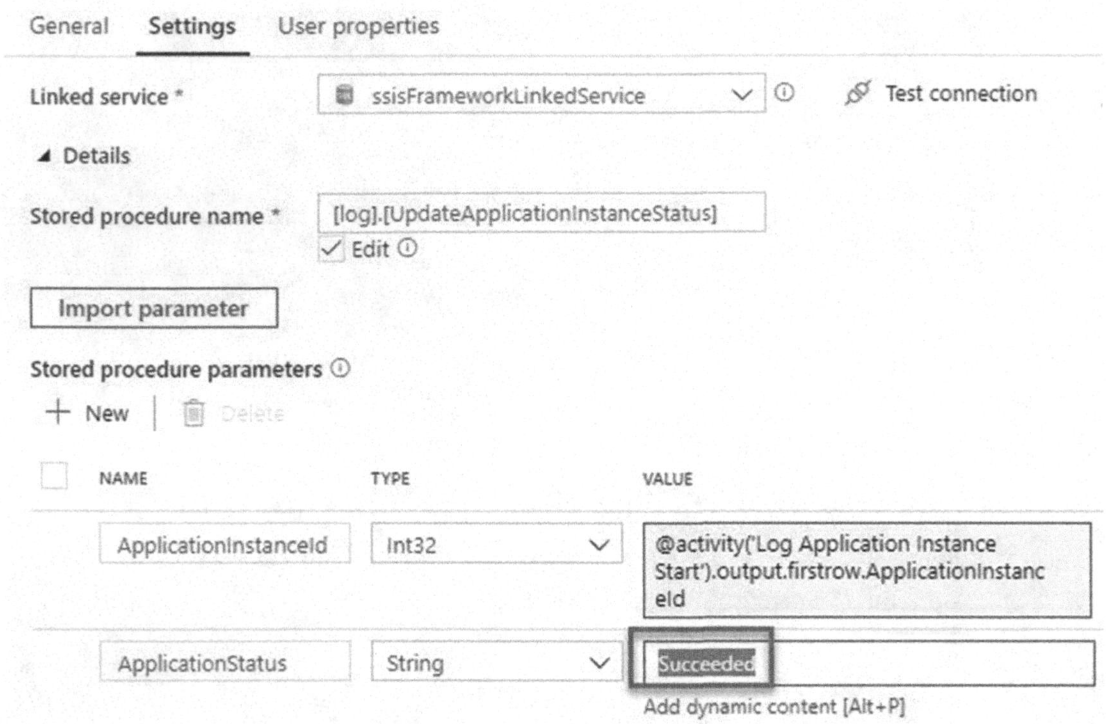

General   **Settings**   User properties

Linked service *   🗟 ssisFrameworkLinkedService   ⌄ ⓘ   ⚹ Test connection

◢ Details

Stored procedure name *   [log].[UpdateApplicationInstanceStatus]
                            ✓ Edit ⓘ

| Import parameter |

Stored procedure parameters ⓘ
  ＋ New | 🗑 Delete

NAME	TYPE		VALUE
ApplicationInstanceId	Int32	⌄	@activity('Log Application Instance Start').output.firstrow.ApplicationInstanceId
ApplicationStatus	String	⌄	Succeeded

Add dynamic content [Alt+P]

Figure 10-21. *Updating the ApplicationStatus parameter value*

A fresh test execution confirms the parent pipeline logs SSIS framework application failures as designed. The Azure Data Factory version of the SSIS framework needs similar instance logging at the application package scope.

The next step is implementing Application Package Instance logging.

Before proceeding, we need to make two changes to the parent pipeline:

- Add the ApplicationPackageId field to the query in the "Get Application Packages" lookup activity.

- Add the ApplicationInstanceId variable (and store the ApplicationInstanceId value in the ApplicationInstanceId variable).

Adding the ApplicationPackageId Field

To add the ApplicationPackageId field to the metadata dataset returned from the SSISConfig database, select the "Get Application Packages" lookup activity, and then click the Settings tab. Modify the Query property using the T-SQL in Listing 10-5.

Listing 10-5. Modified query to retrieve application package metadata

```
Select a.ApplicationName
    , p.PackageLocation + p.PackageName As PackagePath
    , ap.ExecutionOrder
    , ap.FailApplicationOnPackageFailure
    , ap.ApplicationPackageId
From [config].[ApplicationPackages] ap
Join [config].[Applications] a
  On a.ApplicationId = ap.ApplicationId
Join [config].Packages p
  On p.PackageId = ap.PackageId
Where a.ApplicationName = N'Framework Test'
  And ap.ApplicationPackageEnabled = 1
Order By ap.ExecutionOrder
```

After the T-SQL has been modified, the next step is to add a variable to hold the ApplicationInstanceId pipeline variable.

Adding the ApplicationInstanceId Pipeline Variable

To add the ApplicationInstanceId pipeline variable, click the pipeline, and then select the Variables tab. Click the "+ New" button and name the new variable "ApplicationInstanceId" as shown in Figure 10-22.

Parameters	**Variables**	Output	

+ New   🗑 Delete

NAME	TYPE		DEFAULT VALUE
GetApplicationPackages	String	⌄	Value
CancelCommand	String	⌄	Value
ApplicationInstanceId	String	⌄	0

Figure 10-22. *Adding the ApplicationInstanceId pipeline variable*

"Why is the ApplicationInstanceId variable – an integer – declared as a String variable?" That's a fair question. At the time of this writing, the other variable type options are Boolean and Array. The ApplicationInstanceId is neither a Boolean nor an Array.

The next step is to add a Set variable activity to set the value of the ApplicationInstanceId pipeline parameter to the ApplicationInstanceId value returned from the "Log Application Instance Start" lookup activity. To begin, select the Success output of the "Log Application Instance Start" lookup activity, right-click the Success output, and then click Delete, as shown in Figure 10-23.

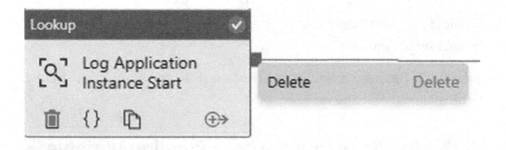

Figure 10-23. *Deleting the Success output from the "Log Application Instance Start" lookup activity*

To add a new Success output to the "Log Application Instance Start" lookup activity, click the "Add Output" icon, and then click "Success," as shown in Figure 10-24.

Figure 10-24. *Adding a new Success output to the "Log Application Instance Start" lookup activity*

Drag a Set variable activity onto the pipeline canvas, and rename it to "Set ApplicationInstanceId." Click the Variables tab, and select ApplicationInstanceId from the Name property drop-down. Click inside the Value property textbox, and then click the "Add dynamic content [Alt + P]" link to open the Set dynamic content blade. Scroll to the Activity outputs category, and click "Log Application Instance Start" to add `@activity('Log Application Instance Start').output` to the expression textbox. Edit `@activity(`, to read `@string(activity(`. Append `.firstrow.ApplicationInstanceId)` to the expression to assign the ApplicationInstanceId value returned in the first row of the output from the "Log Application Instance Start" lookup activity, as shown in Figure 10-25.

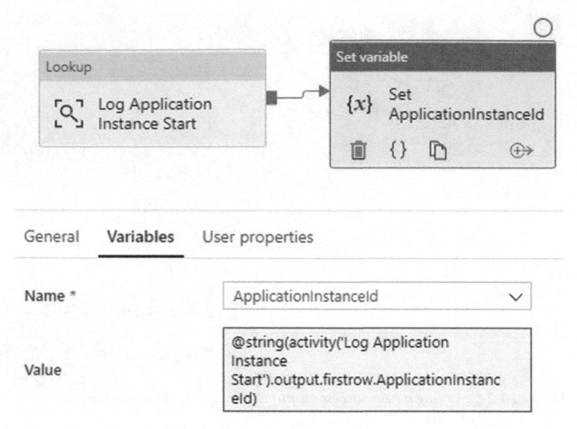

Figure 10-25. *Configuring the "Set ApplicationInstanceId" set variable activity*

Connect a Success output from the "Set ApplicationInstanceId" set variable activity to the "Get Application Packages" lookup activity.

The "Set ApplicationInstanceId" set variable activity is now configured to read the ApplicationInstanceId value returned from the "Log Application Instance Start" lookup activity and store the value in the ApplicationInstanceId pipeline variable.

Finally, update the "Log Application Instance Success" and "Log Application Instance Failure" stored procedure activities to use `@int(variables('Application InstanceId'))` for the ApplicationInstanceId stored procedure parameters instead of the output from the "Log Application Instance Start" lookup activity, as shown in Figure 10-26.

General   **Settings**   User properties

Linked service *   ▣ ssisFrameworkLinkedService   ∨   ⓘ   ⌀ Test connection   ⌀ Open   + New

Stored procedure name *   [log].[UpdateApplicationInstanceStatus]
                          ✓ Edit ⓘ

◢ Stored procedure parameters ⓘ

←⌐ Import   + New | 🗑 Delete

NAME	TYPE	VALUE	
ApplicationInstanceId	Int32 ∨	@int(variables('ApplicationInstanceId'))	
ApplicationStatus	String ∨	Succeeded	☐ Treat as null

Figure 10-26. *Using @variables('ApplicationInstanceId')*

As stated earlier, the ApplicationInstanceId pipeline variable is a string. At the time of this writing, the value would be implicitly converted to an integer here and elsewhere. That said, please be *intentional* in your coding. Here, "intentional" means explicit conversion to an integer value.

The next step is to add a child pipeline to execute the application packages.

Add the Child Pipeline

As mentioned previously, the differences between the on-premises and the Azure Data Factory versions of an SSIS framework drive design differences. The first step in the redesign is migrating application package execution from the parent pipeline to a "child" pipeline.

Begin by creating a new ADF pipeline named "child," as shown in Figure 10-27.

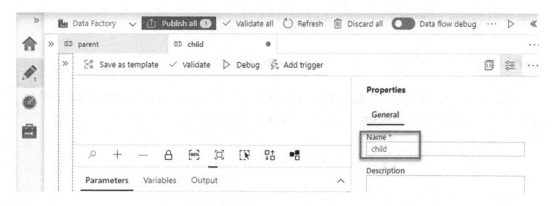

Figure 10-27. *Adding the child pipeline*

In the parent pipeline, the "Execute Application Package" execute SSIS activity has direct access to the properties – populated from SSISConfig metadata – required to dynamically execute application packages related to an SSIS application. The child pipeline will use parameters to access this same metadata. On the Parameters tab of the child pipeline, create the following pipeline parameter / type / default value combinations:

- PackagePath / String / [Empty String]

- ApplicationPackageId / Int / 0

- ApplicationInstanceId / Int / 0

- FailApplicationOnPackageFailure / Bool / true

- ParentRunId / String / [Empty String]

When configured, child pipeline parameters should appear similar to Figure 10-28.

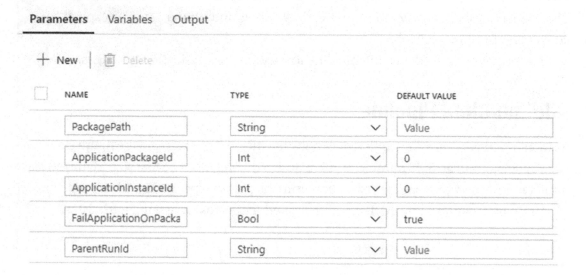

Figure 10-28.   *Child pipeline parameters, configured*

Drag an Execute SSIS package activity onto the child canvas, and rename the new activity to "Execute Application Package," as shown in Figure 10-29.

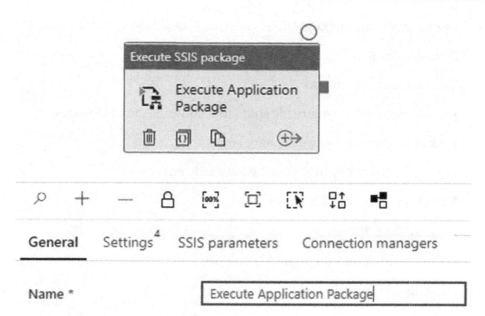

Figure 10-29. *Adding the "Execute Application Package" execute SSIS activity*

Configure the "Execute Application Package" execute SSIS activity's properties on the Settings tab as follows:

- Azure-SSIS IR: Azure-SSIS-Files

- Windows authentication: [Unchecked]

- 32-bit runtime: [Unchecked]

- Package location: File system (Package)

- Package path: @pipeline().parameters.PackagePath

  - Add dynamic content [Alt + P], then parameters ➤ PackagePath.

- Configuration path: [Empty]

- Domain: Azure

- Username: stframeworks

  - The name of the Azure file share that contains the SSIS packages

- Password: [Your file share primary key]

  - The Primary Key to access the Azure file share

- Encryption password: [Empty]

- Logging level: Basic

- Logging path: [Your file share logging path]

- Logging access credentials

  - Same as package access credentials: [Checked if logging folder resides in the same Azure file share as SSIS packages]

When configured, the "Execute Application Package" execute SSIS activity's properties on the Settings tab should appear similar to Figure 10-30.

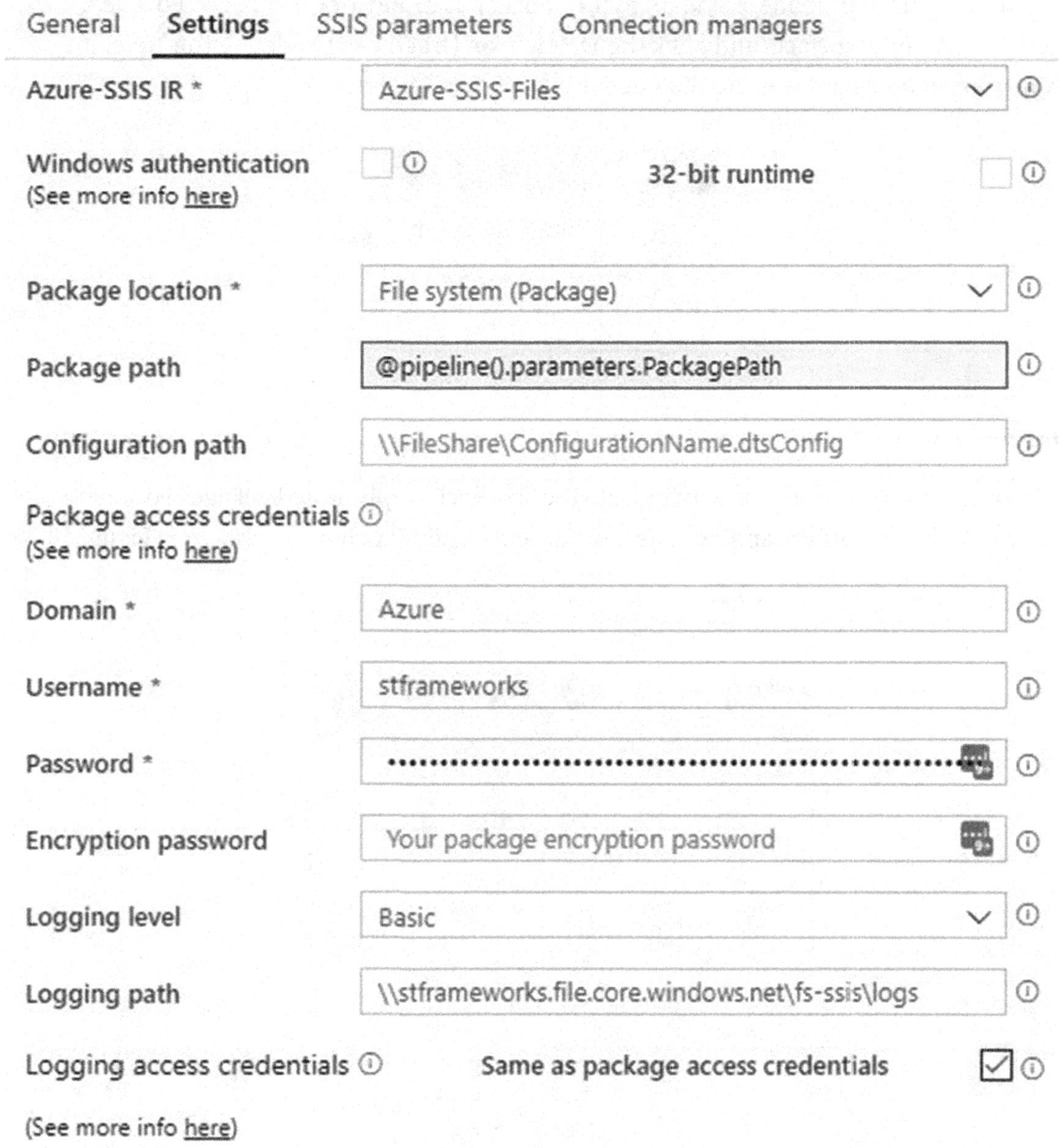

General	**Settings**	SSIS parameters	Connection managers

Azure-SSIS IR *

> Azure-SSIS-Files   ⌄ ⓘ

Windows authentication
(See more info <u>here</u>)   ☐ ⓘ   **32-bit runtime**   ☐ ⓘ

Package location *

> File system (Package)   ⌄ ⓘ

Package path

> @pipeline().parameters.PackagePath   ⓘ

Configuration path

> \\FileShare\ConfigurationName.dtsConfig   ⓘ

Package access credentials ⓘ
(See more info <u>here</u>)

Domain *

> Azure   ⓘ

Username *

> stframeworks   ⓘ

Password *

> ••   ⓘ

Encryption password

> Your package encryption password   ⓘ

Logging level

> Basic   ⌄ ⓘ

Logging path

> \\stframeworks.file.core.windows.net\fs-ssis\logs   ⓘ

Logging access credentials ⓘ   **Same as package access credentials**   ☑ ⓘ

(See more info <u>here</u>)

Figure 10-30. *"Execute Application Package" execute SSIS activity's properties on the Settings tab, configured*

In the parent pipeline, navigate to the "ForEach Application Package" ForEach activity's Activity surface, and click the Delete icon (trash can) to delete the "Execute Application Package" execute SSIS activity, as shown in Figure 10-31.

Figure 10-31. *Deleting the "Execute Application Package" execute SSIS activity*

Drag an Execute Pipeline activity onto the "ForEach Application Package" ForEach activity's Activity surface, and rename it to "Execute Child Pipeline," as shown in Figure 10-32.

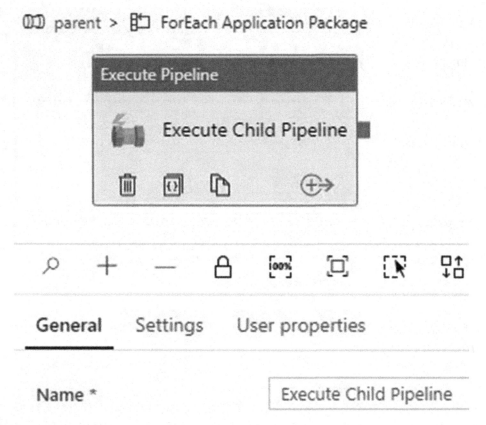

Figure 10-32. *Adding the "Execute Child Pipeline" execute pipeline activity*

Click the "Execute Child Pipeline" execute pipeline activity's Setting tab, and configure the following properties:

- Invoked pipeline: child

- Wait on completion: [Checked]

- Parameters:

  - PackagePath / string / @item().PackagePath

    - Add dynamic content [Alt + P], ForEach iterator ➤ ForEach Application Package, and then append ".PackagePath," as shown in Figure 10-33.

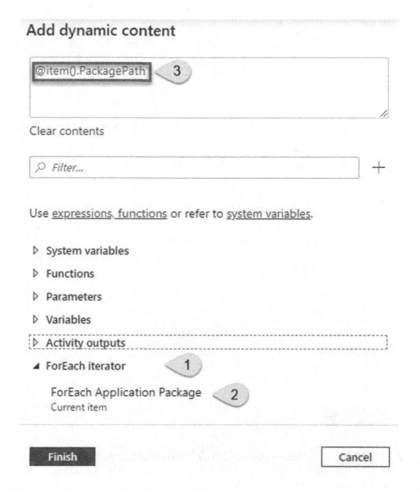

Figure 10-33. *Configuring the PackagePath parameter for the child pipeline*

- ApplicationPackageId / int / @item().ApplicationPackageId

- ApplicationInstanceId / int / @variables('ApplicationInstanceId')

- FailApplicationOnPackageFailure / bool / @item().
  FailApplicationOnPackageFailure

- ParentRunId / string / @pipeline().RunId

    - Add dynamic content [Alt + P], and then System variables ➤
      Pipeline run ID, as shown in Figure 10-34.

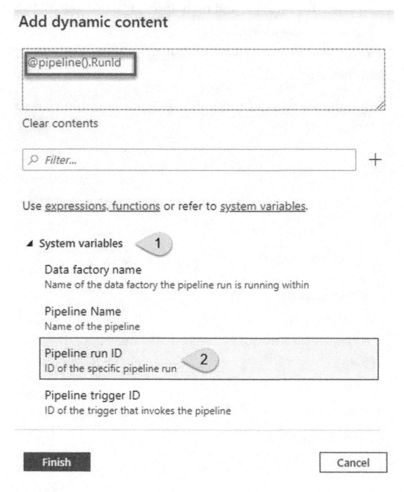

Figure 10-34. Configuring the ParentRunId parameter for the child pipeline

When configured, the "Execute Child Pipeline" execute pipeline activity's properties on the Settings tab should appear similar to Figure 10-35.

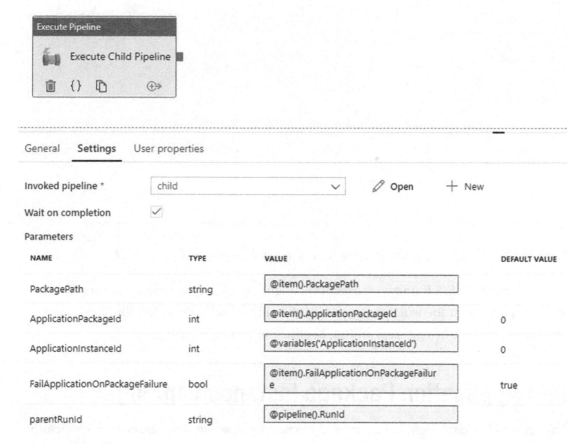

Figure 10-35. *"Execute Child Pipeline" execute pipeline activity's properties on the Settings, configured*

The "Execute Child Pipeline" execute pipeline activity is now configured to call the child pipeline and pass the parameter values required to execute an application package for which metadata is stored in the SSIS framework.

Let's Test It!

Click Debug to test-execute the parent pipeline. The "Execute Child Pipeline" execute pipeline activity should execute twice, failing once and succeeding once, and that's what we see when we view the Output tab of the parent pipeline post-execution, as shown in Figure 10-36.

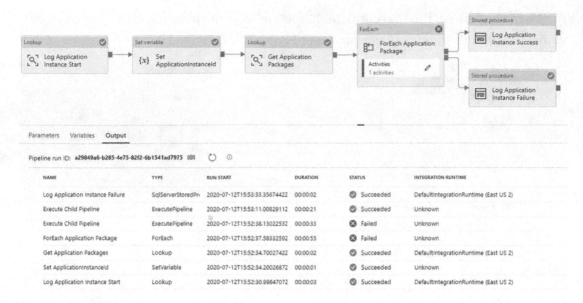

Figure 10-36. *Viewing the parent pipeline's debug execution output*

The updated SSIS framework execution engine works as designed, but there is more to do to implement the fault tolerance for FailApplicationOnPackageFailure, which we will manage after the next step, adding application instance logging.

Add Application Package Instance Logging

Application instance logging happens in the parent pipeline. The pattern for application instance logging is

1. Log application instance start

2. Do some stuff

3. Log application instance status, succeeded or failed

Application instance logging is a solid pattern. Application package instance logging should follow the same pattern.

Modifying the Log.ApplicationPackageInstance Table

As mentioned earlier in the log.ApplicationInstance section, when ADF activities execute, a *RunId* value is generated for use in operational logging. As with logging the parent pipeline's Run Id when an Azure Data Factory parent pipeline executes, logging the child pipeline's RunId is a great idea. The child Run Id value may be used to connect internal ADF pipeline logs to pipeline activity logs.

Modify the log.ApplicationPackageInstance table by executing the T-SQL in Listing 10-6.

Listing 10-6. Adding the ApplicationPackageRunId column to the ApplicationPackageInstance table

```
print 'Log.ApplicationPackageInstance.ApplicationPackageRunId column'
If Not Exists(Select s.[name]
              + '.' + t.[name]
              + '.' + c.[name]
               From [sys].[schemas] s
               Join [sys].[tables] t
                 On t.[schema_id] = s.[schema_id]
               Join [sys].[columns] c
                 On c.[object_id] = t.[object_id]
               Where s.[name] = N'log'
                 And t.[name] = N'ApplicationPackageInstance'
                 And c.[name] = N'ApplicationPackageRunId')
  begin
   print ' - Adding log.ApplicationPackageInstance.ApplicationPackageRunId
   column'
   Alter Table log.ApplicationPackageInstance
    Add ApplicationPackageRunId nvarchar(55) NULL
   print ' - Log.ApplicationPackageInstance.ApplicationPackageRunId column
   added'
  end
```

```
Else
  begin
    print ' - Log.ApplicationPackageInstance.ApplicationPackageRunId column
already exists.'
  end
```

Let's next encapsulate the T-SQL for adding Application Package Instance rows in a stored procedure.

Adding the Log.InsertApplicationPackageInstance Stored Procedure

In the SSIS execution engine (the Parent.dtsx SSIS package), an Execute SQL Task was used to insert the initial ApplicationPackageInstance record into the log. ApplicationPackageInstance table and return an `ApplicationPackageInstanceId` value. In the Azure Data Factory execution engine (*now* the parent *and child* pipelines), a Lookup activity will call a stored procedure that inserts the record and returns an `ApplicationPackageInstanceId` value. The ApplicationPackageInstanceId value will be used later in the pipeline to update the ApplicationPackageInstance record.

In Azure Data Studio (or SSMS), connect to the Azure SQL database named SSISConfig. Starting with the T-SQL code from Listing 7-7 in Chapter 7, edit the T-SQL to create a new stored procedure named "log.InsertApplicationPackageInstance" in an idempotent (re-executable) manner, as shown in Listing 10-7.

Listing 10-7. Idempotent T-SQL to create the log.InsertApplicationPackageInstance stored procedure

```
print 'log.InsertApplicationPackageInstance stored procedure'
If Exists(Select s.[name] + '.' + p.[name]
          From [sys].[procedures] p
          Join [sys].[schemas] s
            On s.[schema_id] = p.[schema_id]
          Where s.[name] = N'log'
            And p.[name] = N'InsertApplicationPackageInstance')
  begin
    print ' - Dropping log.InsertApplicationPackageInstance stored procedure'
    Drop Procedure log.InsertApplicationPackageInstance
```

```
  print ' - Log.InsertApplicationPackageInstance stored procedure dropped'
 end

print ' - Creating log.InsertApplicationPackageInstance stored procedure'
go

Create Procedure log.InsertApplicationPackageInstance
   @ApplicationInstanceId int
 , @ApplicationPackageId int
 , @ApplicationPackageRunId nvarchar(55) = NULL
As

  Insert Into [log].ApplicationPackageInstance
  ( ApplicationInstanceId
  , ApplicationPackageId
  , ApplicationPackageRunId)
  Output inserted.ApplicationPackageInstanceId
  Values
  ( @ApplicationInstanceId
  , @ApplicationPackageId
  , @ApplicationPackageRunId)

go


print ' - Log.InsertApplicationPackageInstance stored procedure created'
go
```

The log.InsertApplicationPackageInstance stored procedure requires the application instance id of the application instance be sent to the @ApplicationInstanceId int parameter in order to *correlate* the application instance with the application package instance. Remember, ApplicationInstanceId is a child pipeline parameter sent to the child pipeline from the parent pipeline. Another child pipeline parameter passed from the parent pipeline is ApplicationPackageId. We pass the value of the ApplicationPackageId pipeline parameter to the log.InsertApplicationPackageInstance stored procedure.

The log.InsertApplicationPackageInstance stored procedure also accepts a parameter named @ApplicationPackageRunId which defaults to NULL when no value is supplied. @ApplicationPackageRunId is used to store the Azure Data Factory child pipeline's RunId value. As stated earlier, storing the RunId in the SSIS framework allows enterprise DevOps teams to relate operational log data in the SSIS framework to operational data captured in ADF logs, which is *extremely* useful when something unfortunate happens during ADF pipeline execution.

The T-SQL `Insert` statement initializes the ApplicationPackageInstance record by inserting new `@ApplicationInstanceId`, `@ApplicationPackageId`, and `@ApplicationPackageRunId` values into the log.ApplicationPackageInstance table. Defaults configured on the log.ApplicationPackageInstance table insert the following values in the row:

- `ApplicationPackageStartTime` is set to `sysdatetimeoffset()` by `DF_log_ApplicationPackageInstance_ApplicationPackageStartTime`.

- `ApplicationPackageStatus` is set to `N'Running'` by `DF_log_ApplicationPackageInstance_ApplicationPackageStatus`.

If all goes as planned, Azure Data Studio should return Messages that appear as shown in Figure 10-37.

```
Messages
─────────

    Started executing query at Line 1
    log.InsertApplicationPackageInstance stored procedure
     - Creating log.InsertApplicationPackageInstance stored procedure
    Started executing query at Line 16
    Commands completed successfully.
    Started executing query at Line 34
     - Log.InsertApplicationPackageInstance stored procedure created
    Total execution time: 00:00:00.136
```

Figure 10-37.   Messages from creating the log.InsertApplicationPackageInstance stored procedure

After the log.InsertApplicationPackageInstance stored procedure has been created, the next step is to add an activity to the pipeline to execute the stored procedure.

Logging Application Package Instance

The ADF execution engine needs to execute a stored procedure, log. InsertApplicationPackageInstance, that returns a value named ApplicationPackageInstanceId. A Lookup activity is built for this!

Return to the child ADF pipeline. Drag a new Lookup activity onto the surface and rename the new Lookup activity to "Log Application Package Instance Start," as shown in Figure 10-38.

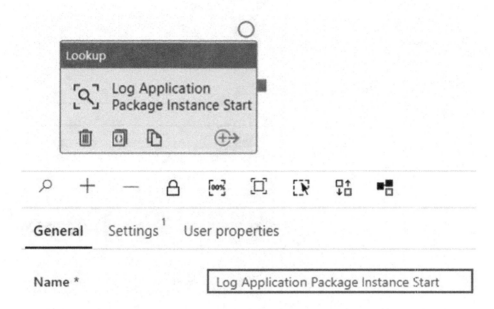

Figure 10-38. *Adding the Log Application Package Instance Start lookup activity*

Click the Log Application Instance Start lookup activity's Settings tab to continue configuration. Click the "Source dataset" property drop-down, and select "ssisFrameworkDataSet," and then set the Log Application Package Instance Start lookup activity's "Use query" property option to "Stored procedure." Select "log. InsertApplicationPackageInstance" from the "Name" property drop-down, as shown in Figure 10-39.

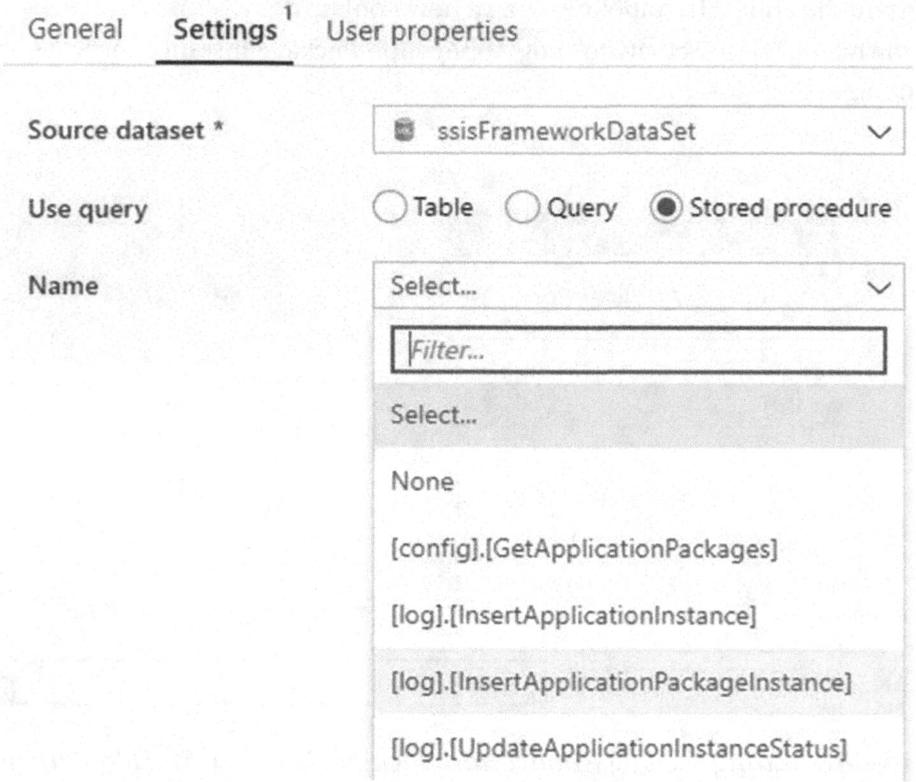

General   **Settings**[1]   User properties

Source dataset *   🗄 ssisFrameworkDataSet   ⌄

Use query   ◯ Table   ◯ Query   ⦿ Stored procedure

Name   Select...   ⌄

| Filter... |

Select...

None

[config].[GetApplicationPackages]

[log].[InsertApplicationInstance]

[log].[InsertApplicationPackageInstance]

[log].[UpdateApplicationInstanceStatus]

Figure 10-39. *Selecting the log.InsertApplicationPackageInstance stored procedure*

Next, click the "Import parameter" button to import the @ApplicationInstanceId, @ApplicationPackageId, and @ApplicationPackageRunId Int32 and String parameters from the log.InsertApplicationPackageInstance stored procedure. Click inside the @ApplicationInstanceId parameter's Value property textbox, and then click the "Add dynamic content [Alt + P]" link, as shown in Figure 10-40.

| Import parameter |

Parameter   + New   🗑 Delete

	NAME	TYPE		VALUE
☐	ApplicationInstanceId	Int32	⌄	Value
				Add dynamic content [Alt+P]
	ApplicationPackageId	Int32	⌄	Value
	ApplicationPackageRu	String	⌄	Value

Figure 10-40. *Preparing to add dynamic content to the parameter value property*

When the "Add dynamic content" blade displays, expand the Parameters expression category and click "ApplicationInstanceId" to set the parameter value ADF expression to @pipeline.parameters.ApplicationInstanceId, as shown in Figure 10-41.

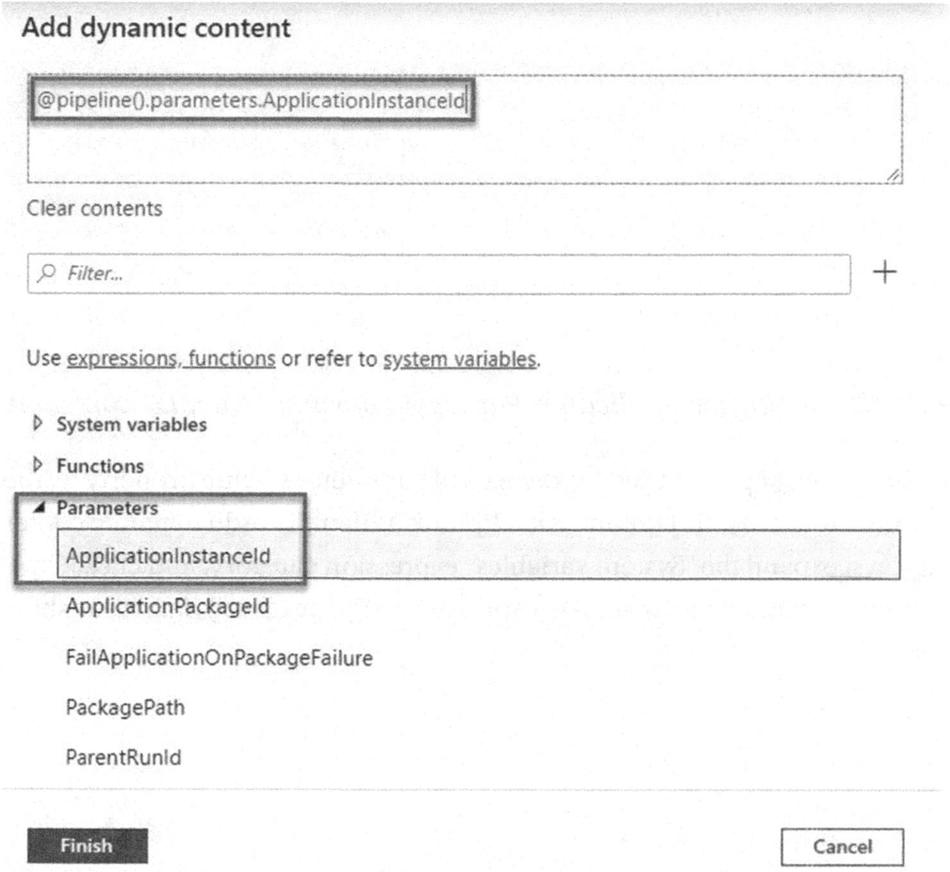

Add dynamic content

@pipeline().parameters.ApplicationInstanceId

Clear contents

🔍 Filter...   +

Use expressions, functions or refer to system variables.

▷ System variables

▷ Functions

◢ Parameters

ApplicationInstanceId

ApplicationPackageId

FailApplicationOnPackageFailure

PackagePath

ParentRunId

Finish   Cancel

Figure 10-41. *Setting the ApplicationInstanceId parameter value expression*

Click the Finish button to proceed.

Repeat the preceding procedure to assign the @ApplicationPackageId stored procedure parameter to the child pipeline's ApplicationPackageId parameter as shown in Figure 10-42.

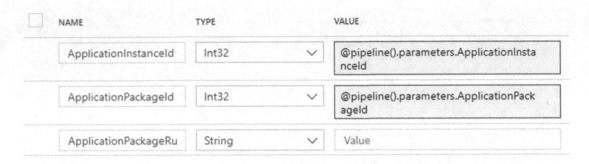

Figure 10-42. *Setting the ApplicationPackageId parameter value expression*

Click inside the @ApplicationPackageRunId parameter's Value property textbox, and then click the "Add dynamic content [Alt + P]" link. When the "Add dynamic content" blade displays, expand the "System variables" expression category, and click "Pipeline run ID" to set the parameter value ADF expression to @pipeline().RunId, as shown in Figure 10-43.

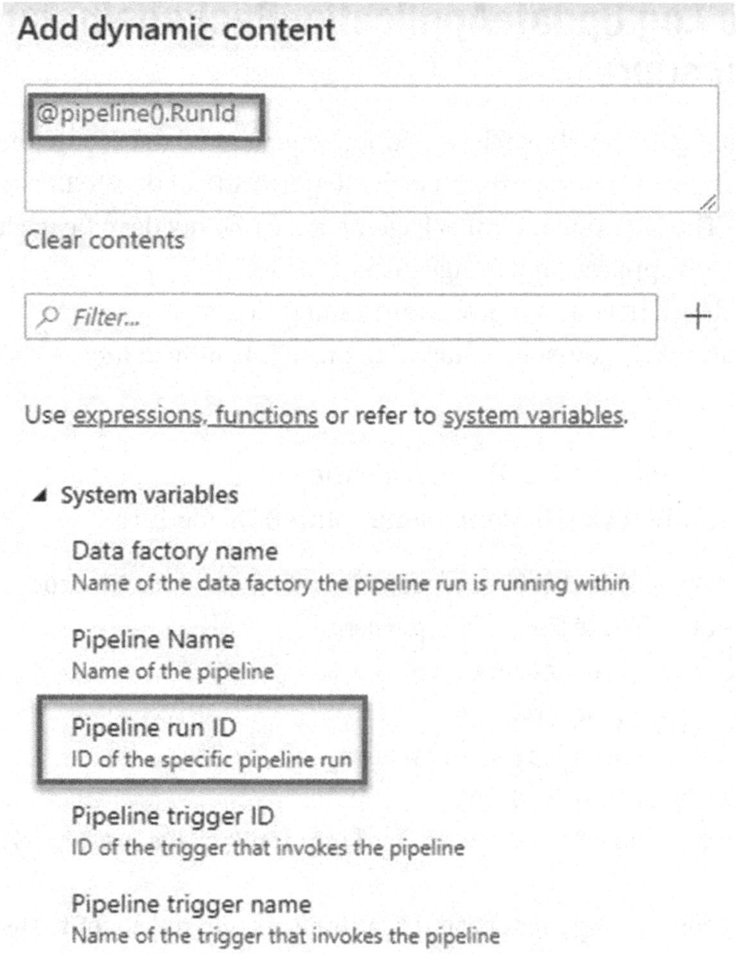

Add dynamic content

@pipeline().RunId

Clear contents

🔍 Filter...   +

Use expressions, functions or refer to system variables.

▲ System variables

Data factory name
Name of the data factory the pipeline run is running within

Pipeline Name
Name of the pipeline

Pipeline run ID
ID of the specific pipeline run

Pipeline trigger ID
ID of the trigger that invokes the pipeline

Pipeline trigger name
Name of the trigger that invokes the pipeline

Figure 10-43. Setting the ApplicationPackageRunId parameter value expression

Leave the "Query timeout (minutes)" property set to its default (120 minutes), the "Isolation level" property set to "None" (default), and the "First row only" property checked (default).

The Log Application Package Instance Start lookup activity is now configured to call the log.InsertApplicationPackageInstance stored procedure, passing the stored procedure the child pipeline's ApplicationInstanceId, ApplicationPackageId, and ApplicationPackageRunId parameter values and returning the ApplicationPackageInstanceId value.

Connect a Success output from the Log Application Package Instance Start lookup activity to the Execute Application Package execute SSIS activity.

Adding the Log.UpdateApplicationPackageInstanceStatus Stored Procedure

As configured, the child pipeline fails execution when called to execute the ReportAndFail.dtsx SSIS package because the ReportAndFail.dtsx SSIS package is designed to fail. The log.ApplicationPackageInstance row needs to be updated to reflect the SSIS framework application package instance failed.

Begin by adding a new stored procedure named "log. UpdateApplicationPackageInstanceStatus" to the SSISConfig database, as shown in Listing 10-8.

Listing 10-8. Idempotent T-SQL to create the log. UpdateApplicationPackageInstanceStatus stored procedure

```
print 'log.UpdateApplicationPackageInstanceStatus stored procedure'
If Exists(Select s.[name] + '.' + p.[name]
          From [sys].[procedures] p
          Join [sys].[schemas] s
            On s.[schema_id] = p.[schema_id]
          Where s.[name] = N'log'
            And p.[name] = N'UpdateApplicationPackageInstanceStatus')
 begin
  print ' - Dropping log.UpdateApplicationPackageInstanceStatus
  stored  procedure'
  Drop Procedure log.UpdateApplicationPackageInstanceStatus
  print ' - Log.UpdateApplicationPackageInstanceStatus stored procedure
  dropped'
 end

print ' - Creating log.UpdateApplicationPackageInstanceStatus stored
procedure'
go

Create Procedure log.UpdateApplicationPackageInstanceStatus
   @ApplicationPackageInstanceId int
 , @ApplicationPackageStatus nvarchar(55) = 'Succeeded'
```

As

```
Update [log].ApplicationPackageInstance
Set ApplicationPackageEndTime = sysdatetimeoffset()
  , ApplicationPackageStatus = @ApplicationPackageStatus
Where ApplicationPackageInstanceId = @ApplicationPackageInstanceId
```

go

```
print ' - Log.UpdateApplicationPackageInstanceStatus stored procedure
created'
go
```

If all goes as planned, Azure Data Studio messages should appear similar to Figure 10-44.

Messages
```
Started executing query at Line 1
log.UpdateApplicationPackageInstanceStatus stored procedure
 - Creating log.UpdateApplicationPackageInstanceStatus stored procedure
Started executing query at Line 16
Commands completed successfully.
Started executing query at Line 28
 - Log.UpdateApplicationPackageInstanceStatus stored procedure created
Total execution time: 00:00:00.114
```

Figure 10-44. Azure Data Studio messages reflecting successful creation of the log. UpdateApplicationPackageInstanceStatus stored procedure

The next step is to add a Stored Procedure Activity that calls the log. UpdateApplicationPackageInstanceStatus stored procedure on failure.

Updating Application Package Instance

The ADF execution engine needs to execute a stored procedure, log. UpdateApplicationPackageInstanceStatus, that accepts ApplicationPackageInstanceId and ApplicationPackageStatus values and then updates the status of the current Application Package Instance. A Stored Procedure Activity is perfect for this!

Return to the child ADF pipeline. Drag a Stored Procedure Activity onto the surface. Add a Failure output from the Execute Application Package activity to the new Stored Procedure Activity. Rename the new Stored Procedure Activity to "Log Application Package Instance Failure," as shown in Figure 10-45.

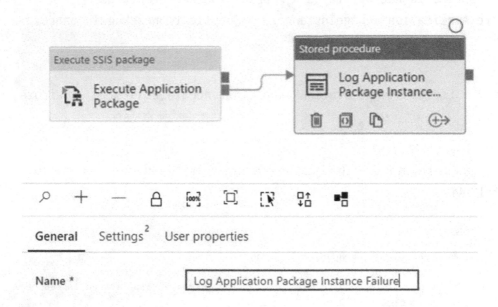

Figure 10-45. Adding the "Log Application Package Instance Failure" stored procedure activity

On the Settings tab, select "ssisFrameworkLinkedService" from the "Linked service" property drop-down, and then select "[log].[UpdateApplicationPackageInstanceStatus]" from the "Stored procedure name" drop-down. Click the "Import parameter" button to import parameters from the log.UpdateApplicationPackageInstanceStatus stored procedure. Click inside the value textbox for the `ApplicationPackageInstanceId` parameter, and then click the "Add dynamic content [Alt + P]" link. When the "Add dynamic content" blade displays, expand the "Activity outputs" expressions category, and click "Log Application Package Instance Start." The dynamic content textbox will display `@activity('Log Application Package Instance Start').output`. Append `.firstrow.ApplicationPackageInstanceId` to the expression to map the value of the `ApplicationPackageInstanceId` returned from the execution of the "Log Application Package Instance Start" lookup activity into the `ApplicationPackageInstanceId` parameter's value property, as shown in Figure 10-46.

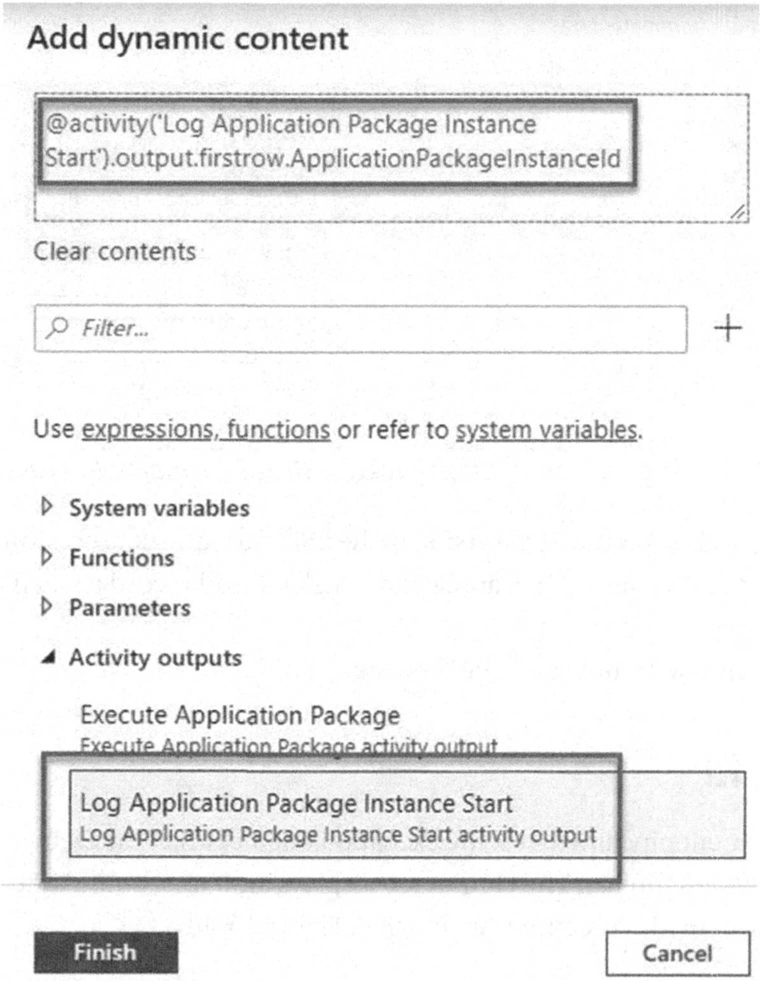

Add dynamic content

@activity('Log Application Package Instance
Start').output.firstrow.ApplicationPackageInstanceId

Clear contents

🔍 *Filter...*   +

Use expressions, functions or refer to system variables.

▷ **System variables**

▷ **Functions**

▷ **Parameters**

◢ **Activity outputs**

   Execute Application Package
   Execute Application Package activity output

   Log Application Package Instance Start
   Log Application Package Instance Start activity output

Finish   Cancel

Figure 10-46. *Mapping the "Log Application Package Instance Start" lookup activity into the ApplicationPackageInstanceId parameter's value property*

Click the Finish button to proceed.

Type "Failed" into the `ApplicationStatus` parameter's value property, as shown in Figure 10-47.

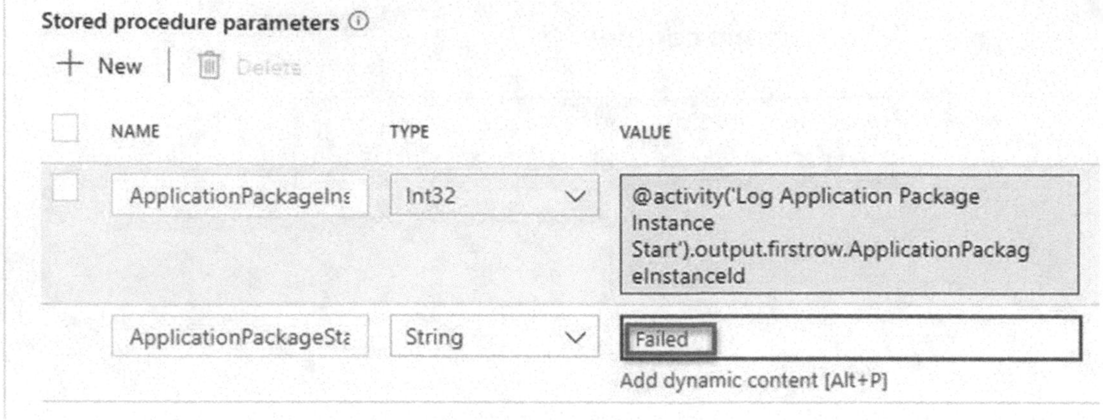

Stored procedure parameters ⓘ

＋ New   🗑 Delete

	NAME	TYPE		VALUE
☐	ApplicationPackageIns	Int32	⌄	@activity('Log Application Package Instance Start').output.firstrow.ApplicationPackageInstanceId
	ApplicationPackageSta	String	⌄	Failed

Add dynamic content [Alt+P]

Figure 10-47. *Setting the ApplicationPackageStatus property to Failed*

The child pipeline for the ADF version of the SSIS framework execution engine is now configured to execute an SSIS application package and record a failed application package instance.

The next step is to test-execute the pipeline.

Let's Test It!

Return to the parent pipeline to test the execution engine. Click the Debug item on the parent pipeline's toolbar. The Output tab displays and, after some time, reveals the parent pipeline activities executed, as shown in Figure 10-48.

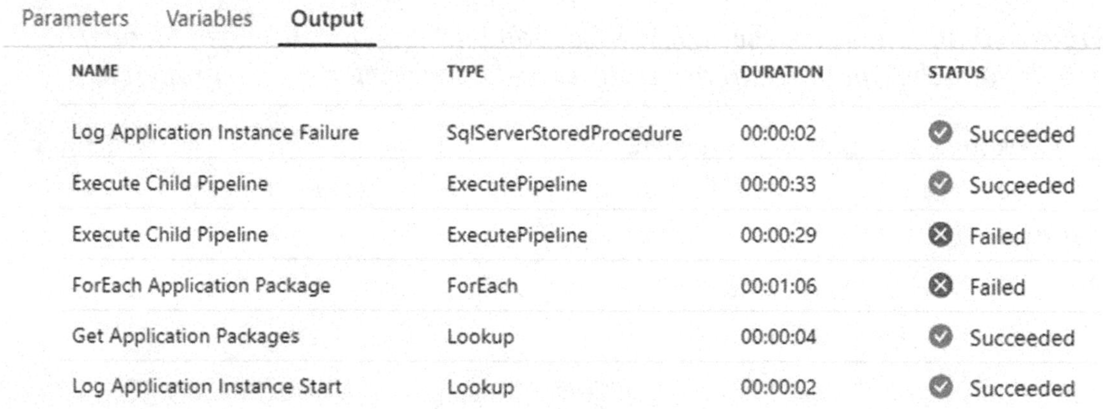

Parameters   Variables   **Output**

NAME	TYPE	DURATION	STATUS
Log Application Instance Failure	SqlServerStoredProcedure	00:00:02	✅ Succeeded
Execute Child Pipeline	ExecutePipeline	00:00:33	✅ Succeeded
Execute Child Pipeline	ExecutePipeline	00:00:29	❌ Failed
ForEach Application Package	ForEach	00:01:06	❌ Failed
Get Application Packages	Lookup	00:00:04	✅ Succeeded
Log Application Instance Start	Lookup	00:00:02	✅ Succeeded

Figure 10-48. *Results of a text execution of the parent pipeline*

Execute the T-SQL query shown in Listing 10-9 to examine the Application Instance results.

Listing 10-9. Application execution results

```
Select top 1
    a.ApplicationName
 , ai.ApplicationInstanceId
 , ai.ApplicationStatus
From log.ApplicationInstance ai
Join config.Applications a
  On a.ApplicationId = ai.ApplicationId
Order By ApplicationInstanceId Desc

declare @ApplicationInstanceId int = (Select top 1 ApplicationInstanceId
                                        From log.ApplicationInstance
                                        Order By ApplicationInstanceId DESC)

Select p.PackageName
 , ai.ApplicationInstanceId
 , api.ApplicationPackageInstanceId
 , api.ApplicationPackageStatus
 , ap.FailApplicationOnPackageFailure
From log.ApplicationPackageInstance api
Join log.ApplicationInstance ai
  On ai.ApplicationInstanceId = api.ApplicationInstanceId
Join config.ApplicationPackages ap
  On ap.ApplicationPackageId = api.ApplicationPackageId
Join config.Packages p
  On p.PackageId = ap.PackageId
Where ai.ApplicationInstanceId = @ApplicationInstanceId
Order By ApplicationPackageInstanceId
```

The results should appear similar to Figure 10-49:

Results   Messages

	ApplicationName	ApplicationInstanceId	ApplicationStatus
1	Framework Test	61	Succeeded

	PackageName	ApplicationInstanceId	ApplicationPackageInstanceId	ApplicationPackageStatus	FailApplicationOnPackageFailure
1	ReportAndFail.dtsx	61	73	Failed	1
2	ReportAndSucceed.dtsx	61	74	Running	1

Figure 10-49. *Application execution results*

While the results match what was expected, something is missing: there is currently no Stored Procedure Activity that updates the Application Package Instance on Success, which is why the results of the query in Listing 10-9, shown in Figure 10-49, display the ReportAndSucceed.dtsx SSIS ApplicationPackageStatus as "Running."

Updating Application Package Instance on Success

The child pipeline needs a way to log Application Package Instance success. Begin by clicking the Clone icon on the "Log Application Package Instance Failure" stored procedure activity as shown in Figure 10-50.

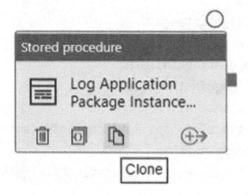

Figure 10-50. *Preparing to clone the "Log Application Package Instance Success" stored procedure activity from the "Log Application Package Instance Failure" stored procedure activity*

As with cloning the "Log Application Instance Failure" stored procedure activity earlier, cloning the "Log Application Package Instance Failure" stored procedure activity makes sense because the new Stored Procedure Activity will be calling the *same* stored procedure – log.UpdateApplicationPackageInstanceStatus – and *only one* parameter value configuration will change.

Connect a Success output from the "Execute Application Package" execute SSIS package activity to the cloned Stored Procedure Activity, and then rename the cloned Stored Procedure Activity to "Log Application Package Instance Success," as shown in Figure 10-51.

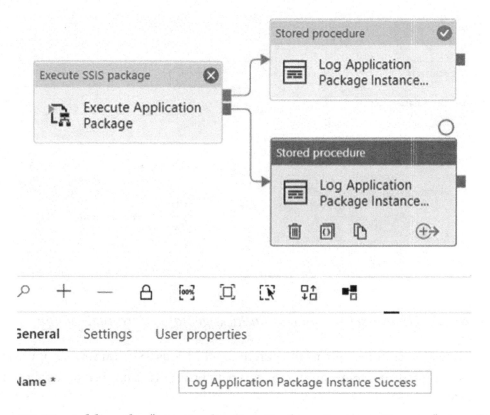

Figure 10-51. *Adding the "Log Application Package Instance Success" stored procedure activity*

Next, click the Settings tab for the "Log Application Package Instance Success" stored procedure activity. The only change required on the Settings tab is the value of the ApplicationPackageStatus parameter's value property, which needs to be updated from "Failed" to "Succeeded," as shown in Figure 10-52.

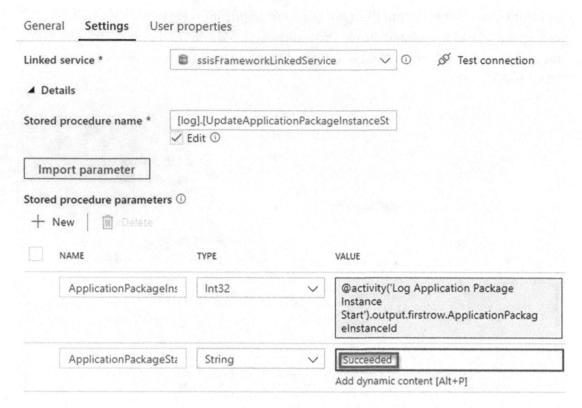

Figure 10-52. *Updating the ApplicationPackageStatus parameter value*

A fresh test execution confirms the parent pipeline logs SSIS framework application failures as designed, and re-executing the query in Listing 10-9 confirms – as shown in Figure 10-53.

	ApplicationName	ApplicationInstanceId	ApplicationStatus
1	Framework Test	62	Running

	PackageName	ApplicationInstanceId	ApplicationPackageInstanceId	ApplicationPackageStatus	FailApplicationOnPackageFailure
1	ReportAndFail.dtsx	62	75	Failed	1
2	ReportAndSucceed.dtsx	62	76	Succeeded	1

Figure 10-53. *ReportAndSucceed.dtsx application package succeeds*

In our examples thus far, each application instance fails because a child package (named "ReportAndFail.dtsx") fails each time it is executed. The next step is to add fault tolerance to the SSIS framework, ADF version, and that is the topic of the next chapter.

Conclusion

In this chapter, we added logging functionality to the parent pipeline, modified the SSISConfig log tables, and encapsulated logging logic in stored procedures. We decoupled the execution of SSIS framework application packages by adding a child pipeline to perform application package execution.

CHAPTER 11

Fault Tolerance in the ADF Framework

The previous chapter covered logging functionality for the Azure Data Factory version of the SSIS framework's execution engine. Differences in functionality available to SSIS packages and functionality available to Azure Data Factory pipelines drove design changes.

In this chapter, we complete ADF execution engine functionality by implementing fault tolerance to programmatically stop pipeline execution based on SSISConfig metadata configurations.

A Brief Introduction to Fault Tolerance

In the author's humble opinion, *fault tolerance* is a different way of describing graceful failure. Thinking "what happens when this fails?" is one difference between a technician and an engineer. Technicians complete projects that accomplish some task, while engineers build solutions that manage issue domains. It comes down to what an individual considers *done*. Technicians work through failures and stop when the project works. Engineers do not stop until the solution does not fail; and – when the solution *does fail* – engineers make certain the failure is graceful.

You may recall fault tolerance was the most complex portion of the design of the on-premises SSIS framework execution engine (the Parent.dtsx package) designed in Chapters 5-7. The design of the Azure Data Factory version of the SSIS framework execution engine (the parent and child pipelines) is no different. To achieve fault tolerance, the ADF execution strategy requires radical redesign.

© Andy Leonard, Kent Bradshaw 2020
A. Leonard and K. Bradshaw, *SQL Server Data Automation Through Frameworks*,
https://doi.org/10.1007/978-1-4842-6213-9_11

The example has already implemented some of the required redesign by building the child pipeline. Why did application package execution need to move to the child pipeline? We answer this question in this chapter.

Add ADF Managed Identity to Contributor Role

In this section, we will use the Azure Data Factory REST API. If that concerns you, don't worry; using the ADF REST API is interesting and will level up your Azure Data Factory automation skills!

The first step is to add the Azure Data Factory Managed Identity to the Contributor role in the Azure Data Factory Access control blade. Begin by logging into the Azure portal and navigating to Azure Data Factory blade. Click "Access control (IAM)," as shown in Figure 11-1.

Figure 11-1. *Navigating to the Access control (IAM) blade*

When the Access control (IAM) blade displays, click the "Add a role assignment" Add button, as shown in Figure 11-2.

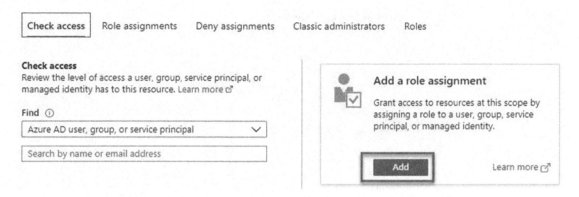

Figure 11-2. *Preparing to assign a role*

When the "Add role assignment" blade displays, select "Contributor" from the Role drop-down. Leave the "Assign access to" drop-down set to the default value ("Azure AD user, group, or service principal"), and enter the name of your Azure Data Factory in the Select textbox, as shown in Figure 11-3.

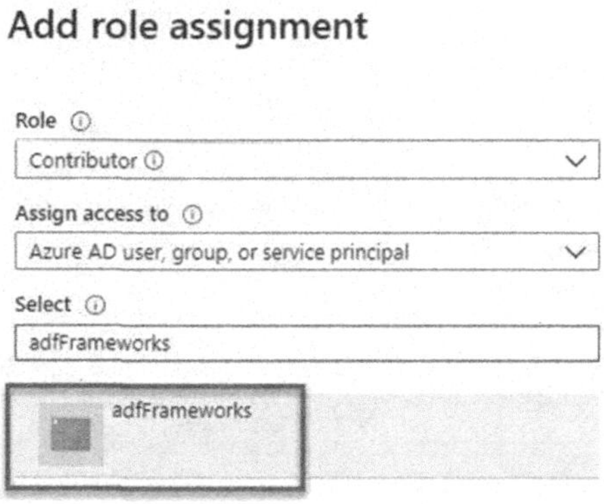

Figure 11-3. *Finding the Azure Data Factory managed identity*

In the "Add role assignment" blade, click the Azure Data Factory managed identity. The "Selected members" list reflects the current selection, and the Save button is enabled indicating the role assignment is ready to store, as shown in Figure 11-4.

Add role assignment

Role ⓘ

| Contributor ⓘ | ⌄ |

Assign access to ⓘ

| Azure AD user, group, or service principal | ⌄ |

Select ⓘ

| adfFrameworks |

| No users, groups, or service principals found. |

Selected members:

adfFrameworks   Remove

Save   Discard

Figure 11-4. *Ready to complete the role assignment*

Click the "Role assignments" tab to view role assignments for ADF, as shown in Figure 11-5.

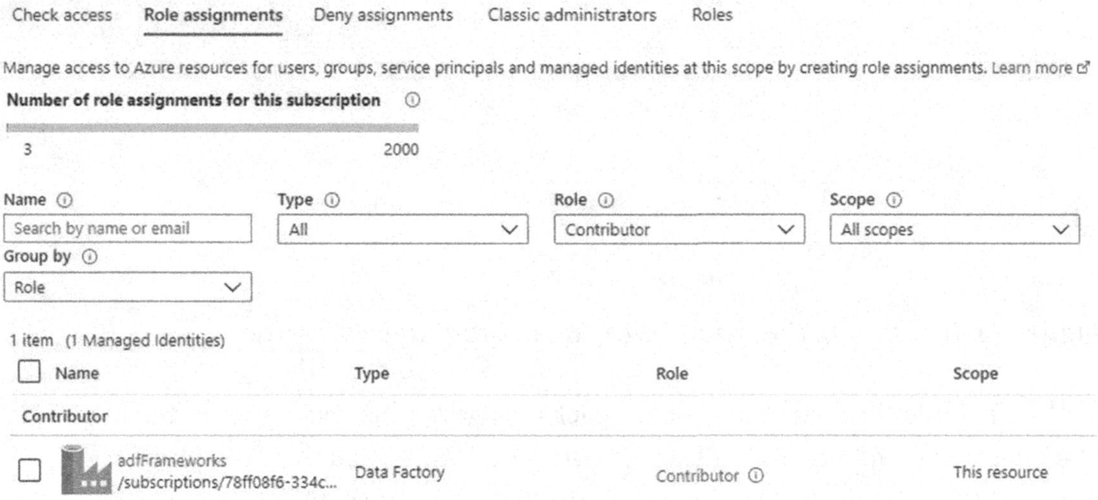

| Check access | Role assignments | Deny assignments | Classic administrators | Roles |

Manage access to Azure resources for users, groups, service principals and managed identities at this scope by creating role assignments. Learn more ⌕

Number of role assignments for this subscription   ⓘ

3   2000

Name ⓘ	Type ⓘ	Role ⓘ	Scope ⓘ
Search by name or email	All ⌄	Contributor ⌄	All scopes ⌄

Group by ⓘ

| Role ⌄ |

1 item (1 Managed Identities)

☐ Name	Type	Role	Scope
Contributor			
☐ adfFrameworks /subscriptions/78ff08f6-334c...	Data Factory	Contributor ⓘ	This resource

Figure 11-5. *ADF managed identity assigned to Contributor role*

Now that the Azure Data Factory managed identity is assigned to the Contributor role, ADF Web activities may interact with the many methods hosted within the Azure Data Factory REST API.

Add Application Package Fault Tolerance

How is fault tolerance implemented within the Azure Data Factory version of the SSIS framework?

About Fail Application on Package Failure

FailApplicationOnPackageFailure is stored along with the application package metadata in the SSISConfig database. The value of FailApplicationOnPackageFailure is returned to the "Get Application Packages" lookup activity in the parent pipeline, and then FailApplicationOnPackageFailure is passed to the child pipeline parameter (named FailApplicationOnPackageFailure) in the "Execute Child Pipeline" execute pipeline activity inside the "ForEach Application Package" foreach activity's inner Activities.

Implementing Application Package Fault Tolerance

Begin implementing application package fault tolerance in the child pipeline by right-clicking the "Execute Application Package" execute SSIS package activity's Failure output and then clicking Delete, as shown in Figure 11-6.

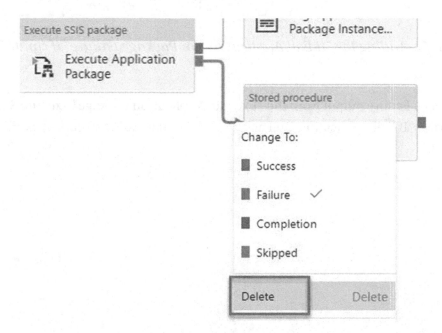

Figure 11-6. *Deleting the "Execute Application Package" execute SSIS package activity's Failure output*

Drag an If Condition activity onto the child pipeline canvas and rename the If Condition "If Fail Application On Package Failure," as shown in Figure 11-7.

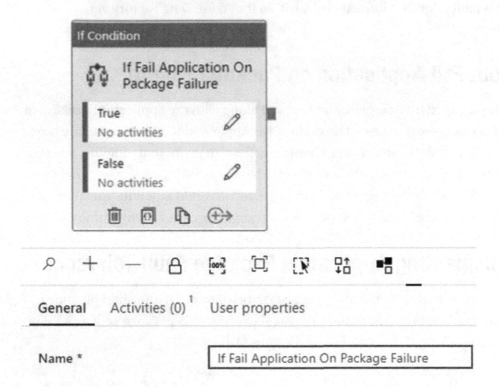

Figure 11-7. *Adding the "If Fail Application On Package Failure" if condition activity*

Connect a Failure output from the "Execute Application Package" execute SSIS activity to the "If Fail Application On Package Failure" if condition activity, as shown in Figure 11-8.

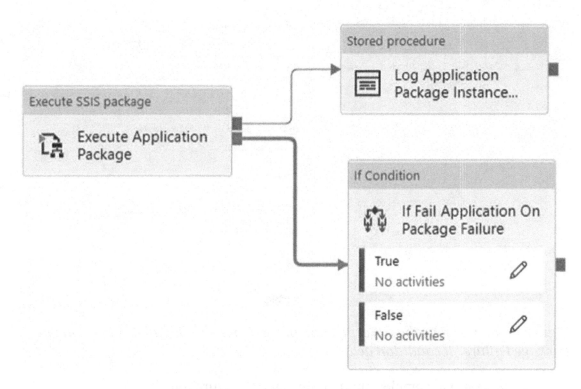

Figure 11-8. *Connecting a Failure output to the "If Fail Application On Package Failure" if condition activity*

Click the "If Fail Application On Package Failure" if condition activity's Activities tab, and then click inside the expression textbox. Click the "Add dynamic content [Alt + P]" link beneath the expression textbox. When the "Add dynamic content" blade displays, expand the Parameters category and select the FailApplicationOnPackageFailure child pipeline parameter. The expression textbox will display "`pipeline().parameters.`
`FailApplicationOnPackageFailure`". Prefix the expression with "`@bool(`", and add a closing parenthesis – "`)`" – to complete the expression, as shown in Figure 11-9.

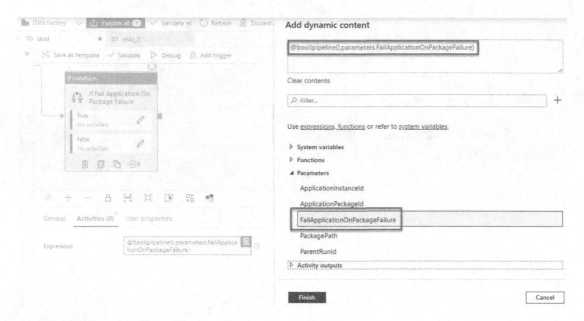

Figure 11-9. *Configuring the expression property for the "If Fail Application On Package Failure" if condition activity*

Click the Finish to complete this portion of the configuration.

Checking the Logic

Let's pause here and think through the logic so far. There are two application package execution possible outcomes: the application package execution will either succeed or fail.

If the application package execution succeeds, the "Log Application Package Instance Success" stored procedure activity executes and updates the log. ApplicationPackageInstance table's `ApplicationPackageStatus` value to "Succeeded" for the current `ApplicationPackageInstanceID`.

The on-success – first use case – is already implemented. Fault tolerance logic is required *if and only if* the application package execution fails.

If the application package execution fails, the "If Fail Application On Package Failure" if condition activity evaluates the value of the child package pipeline's FailApplicationOnPackageFailure parameter (converted by the expression to a Boolean value).

The second use case occurs when the application package execution fails *and* FailApplicationOnPackageFailure is false. In this case, we want to

- Log the status (failure with information) of the application package execution

- *Continue* the execution of the SSIS framework application's application packages

The third use case occurs when the application package execution fails *and* FailApplicationOnPackageFailure is true. In this case, we want to

- Log the status (failure with information) of the application package execution

- Log the status (failure) of the application execution

- *Stop* the execution of application packages

Implementing the Fault Tolerance Logic

To begin implementing the fault tolerance logic, click the "If Fail Application On Package Failure" if condition activity's False activities editor, as shown in Figure 11-10.

Figure 11-10. *Opening the "If Fail Application On Package Failure" if condition activity's False activities editor*

Drag a Stored Procedure Activity onto the canvas and rename the new Stored Procedure Activity "Log Application Package Instance Failed 0," as shown in Figure 11-11.

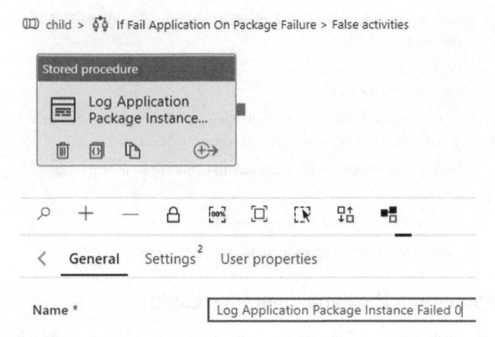

Figure 11-11. *Adding the "Log Application Package Instance Failed 0" stored procedure activity*

Click the Settings tab and set the Linked service property to ssisFrameworkLinkedService. Select log.UpdateApplicationPackageInstanceStatus from the Stored procedure name drop-down, as shown in Figure 11-12.

General	**Settings**	User properties

Linked service *   ssisFrameworkLinkedService   ⌄

◢ Details

Stored procedure name *   [log].[UpdateApplicationPackageInst... ⌄
                               ☐ Edit ⓘ

Figure 11-12. *Configuring Linked service and Stored procedure name*

Click the Import parameter button to begin configuring the log. UpdateApplicationPackageInstanceStatus stored procedure parameter values. Configure the ApplicationPackageInstanceId stored procedure parameter by clicking the "Add dynamic content [Alt + P]" link and selecting the "Log Application Package

Instance Start" activity, which will enter the expression @activity('Log Application Package Instance Start').output into the expression textbox. Append ".firstrow.ApplicationPackageInstanceId" to the expression. In the ApplicationPackageStatus value textbox, enter "Failed (FailAppOnPkgFail: 0)," as shown in Figure 11-13.

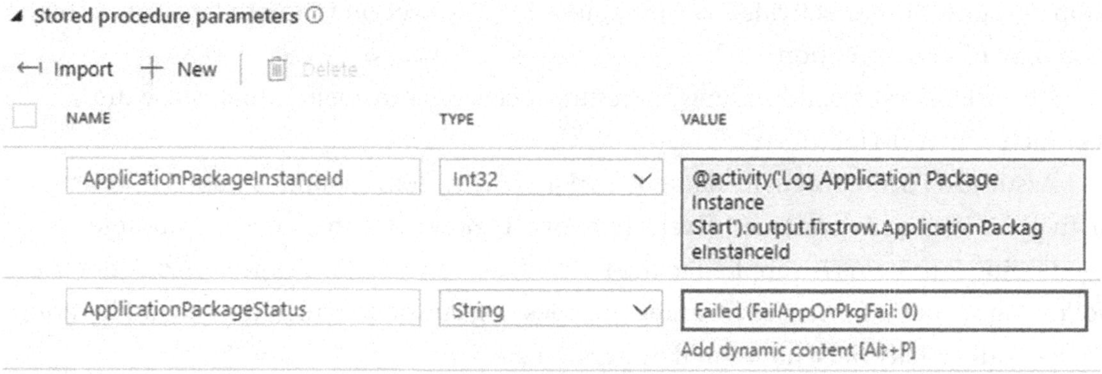

▲ Stored procedure parameters ⓘ

←I Import   + New   🗑 Delete

NAME	TYPE	VALUE
ApplicationPackageInstanceId	Int32 ∨	@activity('Log Application Package Instance Start').output.firstrow.ApplicationPackageInstanceId
ApplicationPackageStatus	String ∨	Failed (FailAppOnPkgFail: 0)

Add dynamic content [Alt+P]

Figure 11-13. *Configuring the ApplicationPackageInstanceId and ApplicationPackageStatus parameter values*

Once stored procedure parameter values are set, configuration of the "Log Application Package Instance Failed 0" stored procedure activity is complete. Please note the ApplicationPackageStatus includes additional information: "(FailAppOnPkgFail: 0)". The "Log Application Package Instance Failed 0" stored procedure addresses use case 2.

The ApplicationPackageStatus column in the log.ApplicationPackageInstance table is currently configured as nvarchar(25), which is not large enough to contain the text "Failed (FailAppOnPkgFail: 0)". Execute the T-SQL in Listing 11-1 to expand the column size.

Listing 11-1. Expanding ApplicationPackageStatus

```
Alter Table [log].[ApplicationPackageInstance]
 Alter Column ApplicationPackageStatus nvarchar(55) Not NULL
```

Before discussing and adding code to support use case 3, a quick review of the ForEach activity's default behavior is a good idea.

ForEach Activity Default Behavior

The ForEach activity's default iterating behavior is to continue iterating over the configured collection, executing "inner activities," until the items in the collection have all been iterated. One side effect is that errors occurring in the "inner activities" do *not* stop iteration. "Inner activities" are executed as the collection is iterated regardless of the outcome of said execution.

After iteration completes, any failures that occurred during iteration cause the ForEach activity to fail.

Assume a ForEach activity named "ForEach Array Item" contains a single "inner activity," a Wait activity named "Wait" (creative, I know). Assume an array variable containing four items is supplied to the ForEach activity's Items property. Consider the following scenario for ForEach "inner activities" execution in which the second iteration of the Wait activity fails, numbered by iteration:

1. Wait executes successfully.

2. Wait execution fails.

3. Wait executes successfully.

4. Wait executes successfully.

Each iteration will execute, resulting in four executions of the wait activity. The ForEach activity will fail. A Debug output will appear as shown in Figure 11-14.

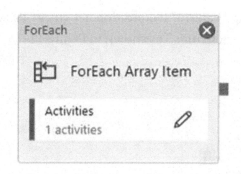

NAME	TYPE	DURATION	STATUS
Wait	Wait	00:00:04	✅ Succeeded
Wait	Wait	00:00:03	✅ Succeeded
Wait	Wait	00:00:00	❌ Failed
Wait	Wait	00:00:02	✅ Succeeded
ForEach Array Item	ForEach	00:00:13	❌ Failed

Figure 11-14. *Output of a ForEach activity execution*

Figure 11-14 shows four executions of the Wait activity. Three succeeded, one failed. Please note each Wait activity executed, and two Wait activity executions fired *after* the Wait activity in the second iteration *failed*. Also note the ForEach activity failed.

ForEach Activity Default Behavior, Applied

Applied to the SSIS framework's fault tolerance, the "ForEach Application Package" ForEach activity – whose "inner activities" are configured in the parent pipeline to execute the child pipeline – is already configured to continue executing the child package regardless of the outcome of the child package execution. The next step, therefore, is to add logic to *stop* application package execution when our use case calls for it – when the application package execution fails *and* FailApplicationOnPackageFailure is true (use case 3).

To configure the "If Fail Application On Package Failure" if condition activity's "True activities," click the Edit icon (pencil) beside "True activities," as shown in Figure 11-15.

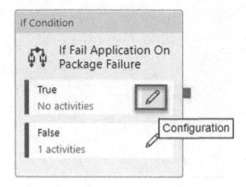

Figure 11-15. *Preparing to configure the "True activities"*

The order of operations for responding to a failed child package execution when the FailApplicationOnPackageFailure bit is configured to true (1, which is the default) is as follows:

1. Log application package instance failure

2. Log application instance cancelled

3. Set the command to cancel the parent pipeline

4. Execute the command to cancel the parent pipeline

The first step for responding to a failed child package execution when the FailApplicationOnPackageFailure bit is configured to true (1) is as follows: log application package instance failure.

To begin adding the functionality to log application package instance failure, add a Stored Procedure Activity to the "If Fail Application On Package Failure" if condition activity's "True activities" canvas and rename the new Stored Procedure Activity "Log Application Package Instance Failed 1," as shown in Figure 11-16.

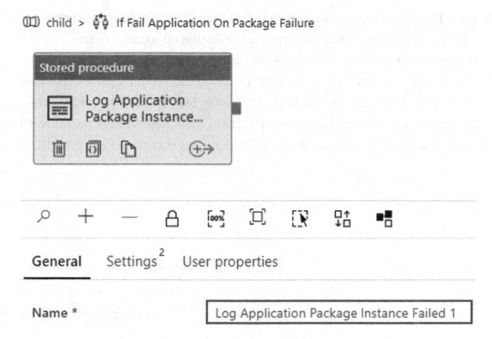

Figure 11-16. *Adding the "Log Application Package Instance Failed 1" stored procedure activity*

On the "Log Application Package Instance Failed 1" stored procedure activity's Settings tab, click the "Linked service" property's drop-down, and select "ssisFrameworkLinkedService". In the "Stored procedure name" drop-down, select "[log].[UpdateApplicationPackageInstanceStatus]," as shown in Figure 11-17.

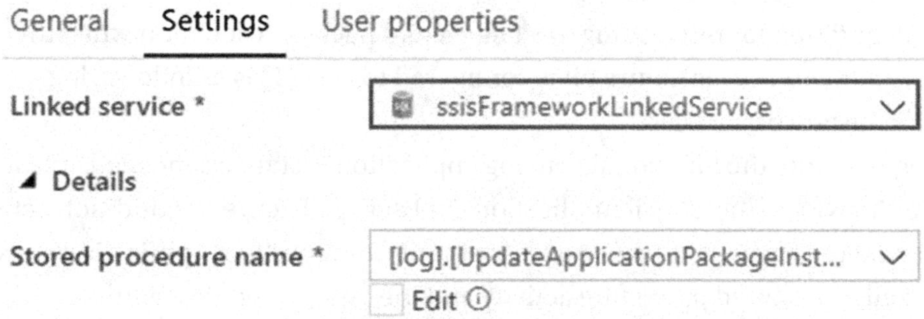

General	**Settings**	User properties

Linked service *	🔲 ssisFrameworkLinkedService	∨

▲ Details

Stored procedure name *	[log].[UpdateApplicationPackageInst...	∨
	☐ Edit ⓘ	

Figure 11-17. *Configuring "Log Application Package Instance Failed 1" stored procedure activity Settings*

Click the Import parameter button. When the `ApplicationPackageInstanceId` and `ApplicationPackageStatus` parameters are displayed, configure the `ApplicationPackageInstanceId` by clicking inside the value textbox, and then clicking the "Add dynamic content [Alt + P]" link, and then entering the expression "@activity('Log Application Package Instance Start').output.firstrow. `ApplicationPackageInstanceId`". In the `ApplicationPackageStatus` parameter value textbox, enter the text "Failed (FailAppOnPkgFail: 1)," as shown in Figure 11-18.

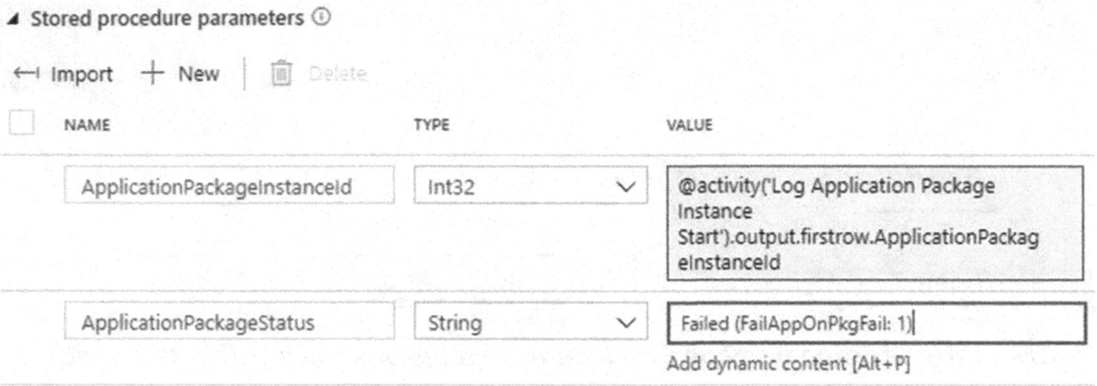

◢ Stored procedure parameters ⓘ

←| Import   ＋ New   ⬚ Delete

NAME	TYPE	VALUE
ApplicationPackageInstanceId	Int32 ⌄	@activity('Log Application Package Instance Start').output.firstrow.ApplicationPackageInstanceId
ApplicationPackageStatus	String ⌄	Failed (FailAppOnPkgFail: 1)

Add dynamic content [Alt+P]

Figure 11-18. *Configuring "Log Application Package Instance Failed 1" stored procedure activity Settings parameter values*

The preceding completes the first step for responding to a failed child package execution when the FailApplicationOnPackageFailure bit is configured to true (1): log application package instance failure.

The second step for responding to a failed child package execution when the FailApplicationOnPackageFailure bit is configured to true (1) is as follows: log application instance cancelled.

To begin adding the functionality to log application instance cancelled, add a Stored Procedure Activity to the "If Fail Application On Package Failure" if condition activity's "True activities" canvas, connect a Success output from the "Log Application Package Instance Failed 1" stored procedure activity and the new Stored Procedure Activity, and then rename the new Stored Procedure Activity "Log Application Instance Cancelled," as shown in Figure 11-19.

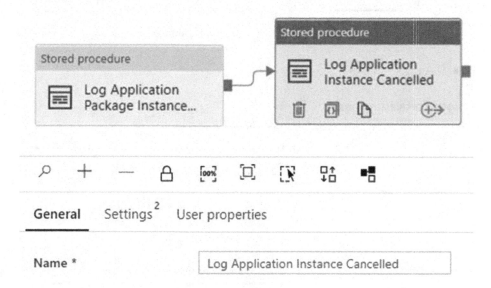

Figure 11-19. *Adding the "Log Application Instance Cancelled" stored procedure activity*

Connect a Success output from the "Log Application Package Instance Failed 1" stored procedure activity to the "Log Application Instance Cancelled" stored procedure activity. On the "Log Application Instance Cancelled" stored procedure activity's Settings tab, click the "Linked service" property's drop-down, and select "ssisFrameworkLinkedService". In the "Stored procedure name" drop-down, select "[log].[UpdateApplicationIntanceStatus]," as shown in Figure 11-20.

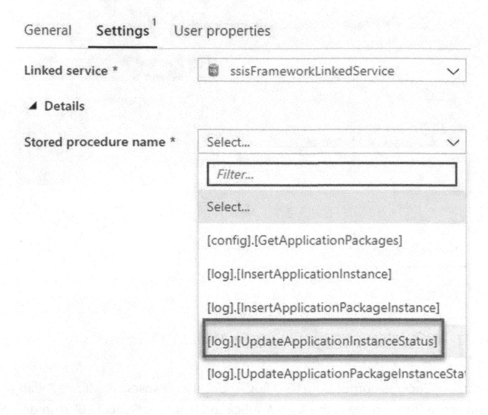

Figure 11-20. *Configuring "Log Application Instance Cancelled" stored procedure activity Settings*

Click the Import parameter button. When the ApplicationInstanceId and ApplicationStatus parameters are displayed, configure the ApplicationInstanceId by clicking inside the value textbox, and then clicking the "Add dynamic content [Alt + P]" link, and then entering the expression "@pipeline().parameters. ApplicationInstanceId". In the ApplicationStatus parameter value textbox, enter the text "Cancelled," as shown in Figure 11-21.

Import parameter

Stored procedure parameters ⓘ

+ New | 🗑 Delete

☐	NAME	TYPE		VALUE	
	ApplicationInstanceId	Int32	⌄	@pipeline().parameters.ApplicationInstanceId	
	ApplicationStatus	String	⌄	Cancelled	

Figure 11-21. *Configuring "Log Application Instance Cancelled" stored procedure activity Settings parameter values*

The preceding completes the second step for responding to a failed child package execution when the FailApplicationOnPackageFailure bit is configured to true (1): log application instance cancelled.

Before we begin this portion of the procedure, navigate to the child pipeline settings, and click the Variables tab. Click the "+ New" button, and add a String variable named "StopParentPipelineRunIdString," as shown in Figure 11-22.

Parameters   **Variables**   Output

+ New | 🗑 Delete

☐	NAME	TYPE		DEFAULT VALUE
	StopParentPipelineRunIdString	String	⌄	Value

Figure 11-22. *Adding the StopParentPipelineRunIdString variable to the child pipeline*

To begin adding the functionality to build the cancel-pipeline run command, add a set variable activity to the "If Fail Application On Package Failure" if condition activity's "True activities" canvas, connect a Success output from the "Log Application Instance Cancelled" stored procedure activity and the new set variable activity, and then rename the new set variable activity "Set Cancel Parent Pipeline Command," as shown in Figure 11-23.

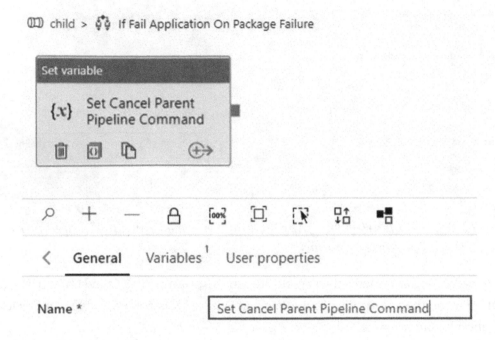

Figure 11-23. *Adding the "Set Cancel Parent Pipeline Command" set variable activity*

On the "Set Cancel Parent Pipeline Command" set variable activity's Variables tab, click the "Name" property's drop-down, and select "StopParentPipelineRunIdString". Click inside the "Value" textbox, and then click the "Add dynamic content [Alt + P]" link. When the "Add dynamic content" blade displays, enter an expression similar to "@concat('https://management.azure.com/subscriptions/<subcription id>/ resourcegroups/<resource group name>/providers/Microsoft.DataFactory/ factories',pipeline().DataFactory,'/pipelineruns/',pipeline().parameters. ParentRunId,'/cancel?api-version=2018-06-01')," as shown in Figure 11-24.

In the expression shown in Figure 11-24, replace <subscription id> with your Subscription Id value and <resource group name> with your Azure Data Factory's Resource Group Name.

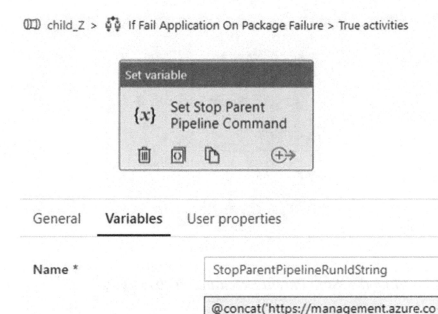

Figure 11-24. *Building the value of the "StopParentPipelineRunIdString"* *variable*

The preceding completes the third step for responding to a failed child package execution when the FailApplicationOnPackageFailure bit is configured to true (1): set the command to cancel the parent pipeline.

The fourth step for responding to a failed child package execution when the FailApplicationOnPackageFailure bit is configured to true (1) is as follows: execute the command to cancel the parent pipeline.

To begin adding the functionality to execute the cancel-pipeline run command, add a web activity to the "If Fail Application On Package Failure" if condition activity's "True activities" canvas, connect a Success output from the "Set Cancel Parent Pipeline Command" set variable activity and the new web activity, and then rename the new web activity "Stop Parent Execution," as shown in Figure 11-25.

Figure 11-25. *Adding the "Stop Parent Execution" web activity*

On the "Stop Parent Execution" web activity's Settings tab, click inside the "URL" property's textbox and click the "Add dynamic content [Alt + P]" link. When the "Add dynamic content" blade displays, enter the expression "@variables('StopParentPipeli neRunIdString')," as shown in Figure 11-26.

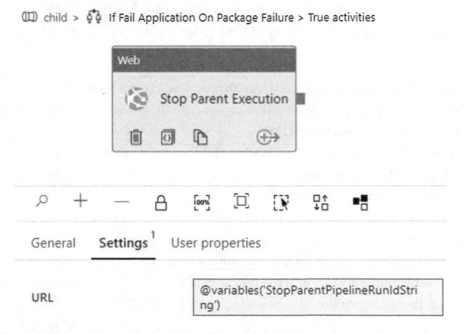

Figure 11-26. *Configuring the URL property of the "Stop Parent Execution" web activity*

Complete configuration of the properties on the "Stop Parent Execution" web activity's Settings tab as follows (and as shown in Figure 11-27):

- Method: POST

- Body: {"message":"Stopping the parent pipeline"}

- Advanced ➤ Authentication: MSI

- Advanced ➤ Resource: https://management.azure.com

General	**Settings**	User properties

URL	@variables('StopParentPipelineRunIdString')
Method *	POST ⌄
Headers	+ New
Body	{"message":"Stopping the parent pipeline"}
Datasets	Select... ⌄
	+ Add dataset reference
Linked services	Select... ⌄
	+ Add linked service reference
Integration runtime	AutoResolveIntegrationRuntime ⌄
◢ Advanced	
Authentication	◯ None ◯ Basic ⓘ ⦿ MSI ⓘ ◯ ClientCertificate ⓘ
Resource *	https://management.azure.com

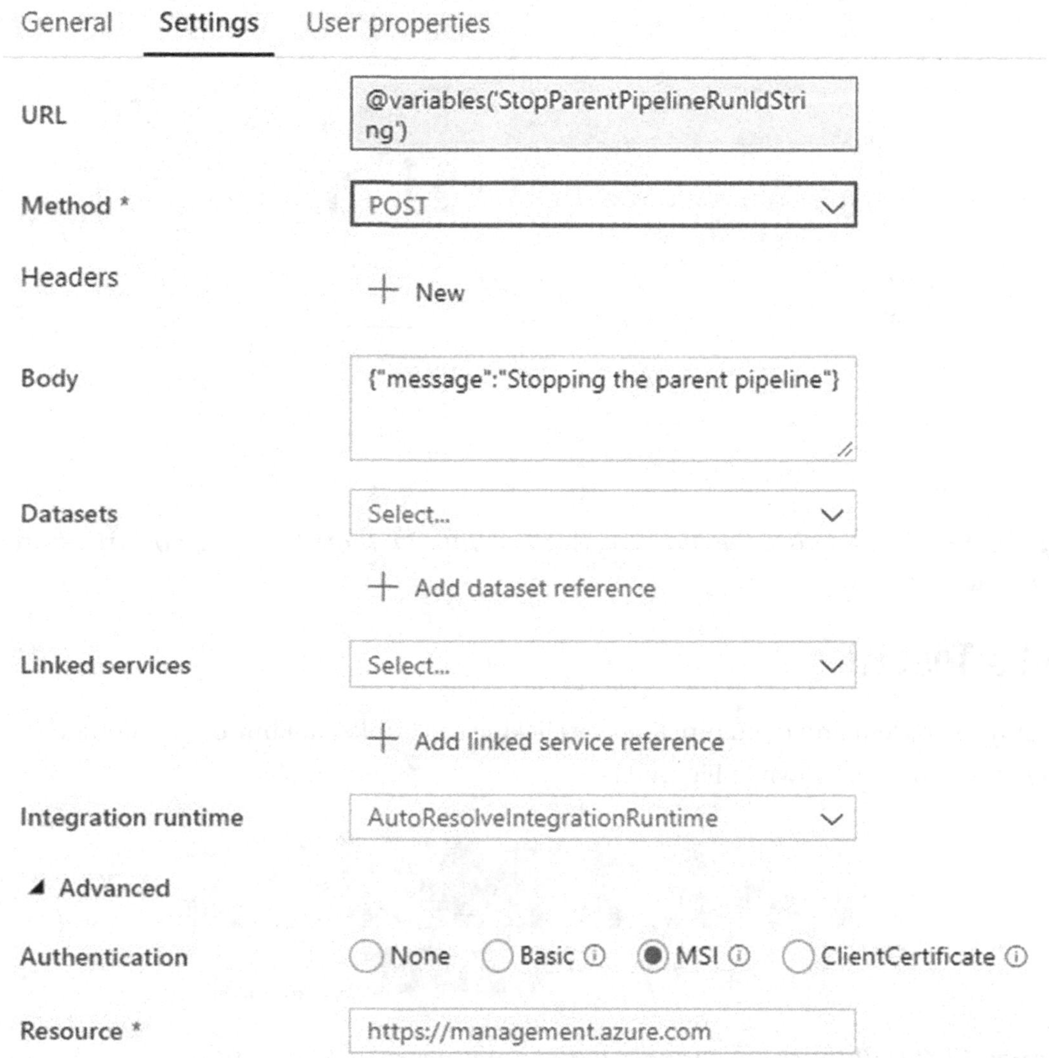

Figure 11-27. *"Stop Parent Execution" web activity Settings configurations*

The preceding completes the fourth and final step for responding to a failed child package execution when the FailApplicationOnPackageFailure bit is configured to true (1): set the command to cancel the parent pipeline.

The activities inside the "If Fail Application On Package Failure" if condition activity's "True activities," as configured, work together to update the statuses of the application package instance and application instance and then invoke the Azure Data Factory's REST API Pipeline Runs Cancel method for the ongoing execution of the parent pipeline.

Before testing the changes, delete the (now orphaned) "Log Application Package Instance Failure" stored procedure activity, as shown in Figure 11-28.

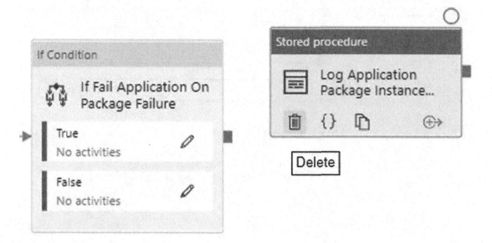

Figure 11-28. *Deleting the old "Log Application Package Instance Failure" stored procedure activity*

Let's Test It!

Deploy the parent and child pipelines by clicking the Publish all button in the Data Factory toolbar, as shown in Figure 11-29.

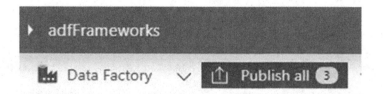

Figure 11-29. *Publishing changes to the pipelines and ADF artifacts*

In the parent pipeline, click the "Add trigger" menu item, and then click "Trigger now," as shown in Figure 11-30.

Figure 11-30. *Triggering the parent pipeline*

When the parent pipeline execution completes, the Monitor page should display results similar to Figure 11-31.

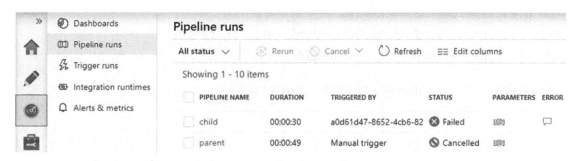

Figure 11-31. *Parent and child pipeline execution results*

Execute the T-SQL query shown in Listing 11-2 to confirm the application instance and application package instance results.

Listing 11-2. Application and application package execution results

```
Select top 1
   a.ApplicationName
 , ai.ApplicationInstanceId
 , ai.ApplicationStatus
From log.ApplicationInstance ai
Join config.Applications a
  On a.ApplicationId = ai.ApplicationId
Order By ApplicationInstanceId Desc

declare @ApplicationInstanceId int = (Select top 1 ApplicationInstanceId
                             From log.ApplicationInstance
                             Order By ApplicationInstanceId DESC)
```

```
Select p.PackageName
  , ai.ApplicationInstanceId
  , api.ApplicationPackageInstanceId
  , api.ApplicationPackageStatus
  , ap.FailApplicationOnPackageFailure
From log.ApplicationPackageInstance api
Join log.ApplicationInstance ai
  On ai.ApplicationInstanceId = api.ApplicationInstanceId
Join config.ApplicationPackages ap
  On ap.ApplicationPackageId = api.ApplicationPackageId
Join config.Packages p
  On p.PackageId = ap.PackageId
Where ai.ApplicationInstanceId = @ApplicationInstanceId
Order By ApplicationPackageInstanceId
```

The results should appear similar to Figure 11-32.

	ApplicationName	ApplicationInstanceId	ApplicationStatus
1	Framework Test	67	Cancelled

	PackageName	ApplicationInstanceId	ApplicationPackageInstanceId	ApplicationPackageStatus	FailApplicationOnPackageFailure
1	ReportAndFail.dtsx	67	82	Failed (FailAppOnPkgFail: 1)	1

Figure 11-32. *Application and application package execution results*

The "Framework Test" SSIS framework application contains two application packages, ReportAndFail.dtsx and ReportAndSucceed.dtsx. As advertised, ReportAndFail.dtsx execution fails every single time it is executed. In the SSIS framework metadata, the FailApplicationOnPackageFailure bit is set to 1 (true) for both application packages, as seen when executing the T-SQL query in Listing 11-3 (with results shown in Figure 11-33).

	ApplicationPackageId	ApplicationName	PackageName	ExecutionOrder	FailApplicationOnPackageFailure
1	2	Framework Test	ReportAndFail.dtsx	10	1
2	1	Framework Test	ReportAndSucceed.dtsx	20	1

Figure 11-33. *Viewing SSIS framework metadata T-SQL query results*

Listing 11-3. Viewing SSIS framework metadata

```
Select ap.ApplicationPackageId
, a.ApplicationName
, p.PackageName
, ap.ExecutionOrder
, ap.FailApplicationOnPackageFailure
From config.Applications a
Join config.ApplicationPackages ap
  On ap.ApplicationId = a.ApplicationId
Join config.Packages p
  On p.PackageId = ap.PackageId
Where a.ApplicationName = N'Framework Test'
Order By ap.ExecutionOrder
```

When the parent pipeline is executed, both application packages – ReportAndFail.
dtsx and ReportAndSucceed.dtsx – are retrieved from the SSISConfig database
and dumped into the parent pipeline's "ForEach Application Package" ForEach
activity's iterator. Because the `ExecutionOrder` of ReportAndFail.dtsx is 10 and the
`ExecutionOrder` of ReportAndSucceed.dtsx is 20, ReportAndFail.dtsx executes first.
By design, when the execution of ReportAndFail.dtsx fails in the child pipeline and the
FailApplicationOnPackageFailure bit is true, the parent pipeline is cancelled.

What happens if ReportAndFail.dtsx fails in the child pipeline and
the FailApplicationOnPackageFailure bit is *false*? To test, update the
FailApplicationOnPackageFailure bit for the ReportAndFail.dtsx application package
using the T-SQL in Listing 11-4.

Listing 11-4. Updating the FailApplicationOnPackageFailure bit for the
ReportAndFail.dtsx application package

```
Select ap.ApplicationPackageId
, a.ApplicationName
, p.PackageName
, ap.ExecutionOrder
, ap.FailApplicationOnPackageFailure
From config.Applications a
Join config.ApplicationPackages ap
```

```
  On ap.ApplicationId = a.ApplicationId
Join config.Packages p
  On p.PackageId = ap.PackageId
Where a.ApplicationName = N'Framework Test'
Order By ap.ExecutionOrder

Update config.ApplicationPackages
Set FailApplicationOnPackageFailure = 0
Where ApplicationPackageId = 2

Select ap.ApplicationPackageId
, a.ApplicationName
, p.PackageName
, ap.ExecutionOrder
, ap.FailApplicationOnPackageFailure
From config.Applications a
Join config.ApplicationPackages ap
  On ap.ApplicationId = a.ApplicationId
Join config.Packages p
  On p.PackageId = ap.PackageId
Where a.ApplicationName = N'Framework Test'
Order By ap.ExecutionOrder
```

The results of executing the T-SQL query in Listing 11-4 should appear similar to Figure 11-34.

Results   Messages

	ApplicationPackageId	ApplicationName	PackageName	ExecutionOrder	FailApplicationOnPackageFailure
1	2	Framework Test	ReportAndFail.dtsx	10	1
2	1	Framework Test	ReportAndSucceed.dtsx	20	1

	ApplicationPackageId	ApplicationName	PackageName	ExecutionOrder	FailApplicationOnPackageFailure
1	2	Framework Test	ReportAndFail.dtsx	10	0
2	1	Framework Test	ReportAndSucceed.dtsx	20	1

Figure 11-34. *FailApplicationOnPackageFailure bit for the ReportAndFail.dtsx application package, updated*

To test the implications of changing the FailApplicationOnPackageFailure bit for the ReportAndFail.dtsx application package, re-trigger the parent pipeline. The Monitor page surfaces the fault-tolerant results: ReportAndFail.dtsx failed but did not stop execution of the SSIS framework application. ReportAndSucceed.dtsx executed and succeeded, as shown in Figure 11-35.

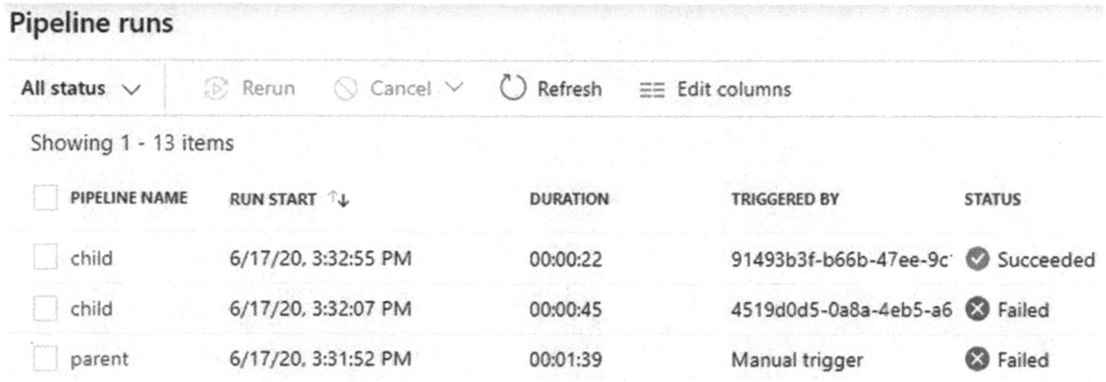

Pipeline runs

All status ∨   ▷ Rerun   ⃠ Cancel ∨   ↻ Refresh   ≣ Edit columns

Showing 1 - 13 items

	PIPELINE NAME	RUN START ↑↓	DURATION	TRIGGERED BY	STATUS
☐	child	6/17/20, 3:32:55 PM	00:00:22	91493b3f-b66b-47ee-9c·	✅ Succeeded
☐	child	6/17/20, 3:32:07 PM	00:00:45	4519d0d5-0a8a-4eb5-a6	❌ Failed
☐	parent	6/17/20, 3:31:52 PM	00:01:39	Manual trigger	❌ Failed

Figure 11-35. *Fault tolerance in action; an application package failure does not stop application execution*

My only complaint is that the parent pipeline reports failure.

Conclusion

In this chapter, we added fault tolerance to the Azure Data Factory version of the SSIS framework. In the process, we learned about configuring security for the Azure Data Factory Managed Identity so web activities in the data factory's pipelines could invoke methods in the Azure Data Factory REST API. We also learned more about the default behavior of the ForEach activity.

Index

A, B

Azure Data Factory (ADF) framework, 209
  IAM
    add role assignment, 357
    complete task, 358
    Contributor role, 358
    Select textbox, 357
Azure-SSIS framework
  Azure Storage Explorer
    create File Share, 269
    explorer window, 268
    fs-ssis, 269, 270
    install, 267
    stframeworks storage account, 268
    update PackageLocation Values, 272
    upload SSIS packages, 270
  SSISConfig database
    complete portal, 249
    "Compute + storage" property, 244
    configure database, 245
    connection details, 250
    create Database, 242
    create SSISConfig database, 251
    data details, 244
    enable advanced data security property, 247
    reconfiguring database, 246
    results of, query execution, 264
    Review + create page, 248
    SSISConfig DDL, 252–264

    subscriptions and resource group property, 243
    T-SQL query, 267
    updated settings, 247
Azure-SSIS integration
  ADF
    Azure dashboard, 215
    Azure Data Factory page, 216
    create button to provision, 215
    create resource, 209
    Data Factory Overview page, 216
    data factory page, 210
    enable GIT checkbox, 214
    new data factory page, 211
    resource group drop-down, 212
    searching data factory, 209
    select location, 213
  Azure Storage
    Access tier property, 220
    configure account kind property, 219
    configure replication property, 220
    configuring storage account, 218
    create storage account, 217, 221, 223
    instance details, 219
    performance property, 219
    storage account name, 218
    storage account setting, 222

385

© Andy Leonard, Kent Bradshaw 2020
A. Leonard and K. Bradshaw, *SQL Server Data Automation Through Frameworks*,
https://doi.org/10.1007/978-1-4842-6213-9

Printed in the United States
By Bookmasters